Making It

Manufacturing Techniques for Product Design

Laurence King Publishing

Contents

For Jerome, our treasure

First published in Great Britain in 2007.
Second edition published 2012 by Laurence King Publishing in association with Central Saint Martins College of Art & Design

Reprinted 2013

This book has been produced by
Central Saint Martins Book Creation,
Southampton Row, London WC1B 4AP, UK

Laurence King Publishing Ltd
361–373 City Road
London EC1V 1LR
United Kingdom
Tel: + 44 20 7841 6900
Fax: + 44 20 7841 6910
e-mail: enquiries@laurenceking.com
www.laurenceking.com

A catalogue record for this book is available from the British Library

ISBN: 978-1-85669-749-1

Design: Roger Fawcett-Tang, Struktur Design
Cover design: Marianne Noble
Senior editors: Peter Jones,
Jessica Spencer (2007 edition)
Picture research: Jennifer Hudson, Lucy Macmillan
Printed in China

Making
It
440369

4 Thin & Hollow

5 Into Solid

6 Complex

7 Advanced

8 Finishing Techniques

Introduction

We are drawn to the unknown, to uncovering secrets and to unearthing the unseen nature of the modern world – from childhood TV shows that peer through the windows of factories looking at lines of chocolate biscuits and milk bottles being made, to cottage industries that reveal to tourists the production methods of indigenous craftsman, even down to DVD 'bonus features' that entertain us with how filmmakers cheat reality with special effects. Designers in particular are constantly looking for new ways to transform both old and new technologies and to apply them within the design arena.

The invention of machines that are used to turn out two thousand light bulbs per minute or ultra-fine flexible fibre optic cables has always amazed me and how they came into being. What kind of creative mind would have conceived the process that requires hot, sticky, molten glass to be suspended from a tower and dropped at a slow rate, stretching into tubes of glass less than 1mm thick to create an optical glass fibre; or that contorts steel wire into the iconic Gem paper clip at a rate of 300 per minute; or the process that makes glass marbles with those swirly patterns of colour, each with its own pattern? But these uniquely formed products are each specific to their own production process and are for another book. Instead this book presents those methods that are relevant to the production of any given object, in a nutshell those that are relevant to industrial design.

Since the publication of the highly successful first edition of *Making It* there have been many innovations in the world of production, a selection of which has been added to this new updated edition. Some are highly specialized, such as electromagnetic forming, some are old methods re-evaluated, such as Industrial Origami®, and others combine two processes, such as Exjection®, a method of making that combines injection moulding and extrusion. There have also been some notable uses of established production methods, such as Marcel Wanders' Sparkling Chair, which upscales injection blow molding from the plastic bottle industry and applies it to the making of a piece of furniture.

There is a growing momentum for a more sustainable approach to design. In order to address the increasing importance of energy use, material scarcity and ethical production I have also added details that will give the reader an introduction to some of these complex areas and key points of consideration. In addition there is also a major new section on finishing techniques, as one of the most common ways is to innovate by colouring, painting, spraying, growing or adding functionality to a component. The chart on pages 10 to13 provides an at-a-glance overview of major factors such as volumes of production and cost for each technique that will give the reader a straightforward comparison and quick reference guide to these key pieces of information.

As stated, the aim of this book is to explore the hidden side of objects within the context of industrial design. To peer into the world of machines and the often creative and inspired ways that they have been assembled to morph liquids, solids, sheets, powders and hunks of metal into three-dimensional products. To examine these pieces of theatre in a way that has not been presented before in a book. To try and communicate what is an inherent playfulness that is evident in mass production. To encourage the abduction of some of these methods by designers to make better products and to exploit production as part of the creative process rather than as a means to an end.

My intention was to take all the information that exists in technical manuals, trade journals and websites for associations and federations in the engineering sectors and to encapsulate it in a guide for the designer that would be a relevant introduction to the world of the manufactured object. In one sense to celebrate all the relevant methods of mass- and batch production for the three-dimensional object at this particularly important crossroads in the evolution of objects. This is a time when old ideas of manufacturing are being re-evaluated by the design industry and new possibilities are surfacing, which have the potential to alter dramatically the way we make, choose and consume our products. It used to be the case that design was the slave to manufacturing, restricting creativity, moulding constraints and costs. In many cases this is still true but increasingly manufacturing is seen as a tool for designing new opportunities to bring new materials and ideas to new methods of production, and to experiment with preconceived volumes of production.

Some of the examples featured in this book reflect a stage of development where the new tools of designers and makers are not physical tools but factory set-ups. Take, for example, Malcolm Jordan's Curvy Composites, a degree-show design project that resulted in a completely new way of forming wood. I couldn't resist also including some of the more offbeat processes that perhaps don't really fit into the realm of the mass-produced object but help point the way to a new direction, ideas that take a type of industrial production and combine it with a craft-like approach, projects that take small-scale, widely available machines and re-use them.

Before the Industrial Revolution the crafting of objects was often influenced by surrounding geography. Ceramics, for example, were designed and made in areas where there was an abundance of clay, such as Stoke-On-Trent in the north-west of England, the birthplace of Wedgwood and countless other ceramics factories. Places with large areas of woodland often had communities that specialized in furniture production. Skills and materials came from specific local resources. The global economy has had consequences for local resources and has often destroyed communities, but now technology is taking production back to the small-scale craft user and placing it in the hands of the consumer. Sometimes this is intentional and driven by new products and technology, at other times it is driven by people abducting machines and using them for something for which they were not intended. The adaptation of the humble inkjet printer for rapid prototyping is one such example.

The re-use of this type of product or technology is a vital part of evolution: experimenting, mixing things up and swapping them around, turning existing conventions on their head. Our insatiable appetite for making things races ahead at full speed. But if the old tools of craftsmanship were hand tools for shaping materials then the new tools of the craftsman are the machines. For under £100 you can buy an inkjet printer, take out the guts and start playing with the workings and use CAD-driven data to produce a whole range of new things. When people first started making 'things' they picked a lump of wood, understood its properties to a certain degree and were able to chop it into a useable product. For some, the lump of wood has become the inkjet printer, a piece of technology that has been chopped up and generally messed about with to create a multitude of products.

Possibly one of the most unusual technologies of this kind is that which has been developed by various teams of scientists across the world using 'modified inkjet printers' to build up living tissue. Based on the long-held knowledge that when placed next to one another, cells will weld together, the tissue is built up using a thermo-reversible gel as a kind of scaffolding over each cell. A team from Medical University of South Carolina uses this gel as a way to support the cells as they are being distributed. This gel is interesting in itself, designed to change instantly from liquid to gel in response to a stimulus such as increase in temperature. This would allow tissue to be placed inside the body supported by the gel; the gel would then dissolve.

The core of this book, however, deals with mass-production techniques, some well established, others very new. In order for these tools to be used, they need to be understood in all their forms and to be presented in a manner that is relevant to design, stimulating ideas and allowing for new creative connections to be made, connections that could provoke the re-appropriation of a technology into a new area or industry. The structure and layout of this book are straightforward and allow for a casual toe-dipping into the world of the manufactured object, hopefully to inform and inspire a fresh look and a greater understanding of the backstage arena of the world of consumerism.

How to use this book

The book is divided into sections based on the shapes of components that can be produced with each process. It does not set out to answer all the questions you might ever need to know about these methods but does provide a clear and basic introduction, using a combination of text, illustrations and photographs of the products. The visual explanations provided in the diagrams serve to encapsulate the principles of the process and the steps that go into making a final component. They are not meant to be accurate drawings of the machines.

The text for each feature is broken down into two main forms to provide a summary of the particular process and a secondary list of the key points that relate to the process.

Pros and cons

These are bullet-pointed notes that summarize each production method to provide a quick guide to key features.

Volumes of production

This explains the range of unit volumes that different methods are capable of, from one-off rapid prototyping to single production runs in the hundreds of thousands.

Unit price vs capital investment

One of the main criteria for specifying a particular method of production is knowing the initial investment that is required. This can vary enormously from plastic forming methods, such as the various forms of injection moulding, which potentially can run into tens of thousands of pounds, to CAD-driven methods, which require no tooling and minimal set-up costs.

Speed

Speed is an important factor in understanding the scale of production and how many units can be produced over a period of time. If, for example, you want to make 10,000 glass bottles, then the automated glass blow moulding process, which can produce 5,000 pieces per hour, is not for you, as the set-up and tooling would prohibit such a short production time on the machines.

Surface

This briefly describes the type of surface finish that you can expect from a particular process. Again this can vary enormously and indicates whether a part would need a secondary process in order to arrive at a finished part.

Types/complexity of shape

This offers guidance on any restrictions that will affect the shape of the component and any design details to consider.

Scale

This gives an indication of the scale of the products that can be produced from the particular process. Sometimes this can offer some surprising facts, for example some metal spinners can spin metal sheets up to 3.5 metres in diameter.

Tolerances

The degree of accuracy that a process is capable of achieving is often determined by the material. Machine-cut metals or injection-moulded plastics, for example, are capable of highly controlled tolerances. Certain ceramic processes, on the other hand, are much less able to achieve precise finished dimensions. This section gives examples of this accuracy.

Relevant materials

This is a list of the types and range of materials that can be formed with the featured process.

Typical applications

A list of products and industries that typically utilize the method of production, the word 'typically' has to be emphasized as the list is not exhaustive but gives sufficient examples to help explain the process.

Similar methods

This provides a key to other processes featured in the book that might be looked at as an alternative form of production to the one featured.

Sustainability issues

A very brief overview of some of the key points that fall into the area of sustainability. This will allow the reader to make more informed decisions on areas such as energy use, toxic chemicals and material wastage.

Further information

This lists web resources to visit for further information. These include contributors to the book. Any relevant associations, where available, are also listed.

Comparing Processes

The chart that runs over the next few pages will enable you to compare different processes and evaluate which is best for your own product. Processes are listed according to chapter type and in the same order as they appear in the relevant chapter. See the chapter entry for further detail.

Key: ★= low, ★★= mid, ★★★= high	Cost of capital investment	Number of components produced per hour	Quality of surface finish
1. Cut from Solid			
Machining	★	★ / ★★	★★★
Computer Numerical Controlled (CNC) Cutting	★	★ / ★★	★★★
Electron-Beam Machining (EBM)	★★	★	★
Turning	★	★ / ★★	★★
Jiggering and Jollying	★	★★ / ★★★	★★★
Plasma-Arc Cutting	★	★★ / ★★★	★★★ (edge surface finish)
2. Sheet			
Chemical Milling	★	★★★	★★
Die Cutting	★★	★★	★★ (edge surface finish)
Water-Jet Cutting	★	★ / ★★	★★ (edge surface finish)
Wire Electrical Discharge Machining (EDM)	★	★ / ★★	★★
Laser Cutting	★	★ / ★★	★★ (wood) / ★★★ (metal)
Oxyacetylene Cutting	★	★★	★★
Sheet-Metal Forming	★★ / ★★★	★ / ★★	★
Slumping Glass	★★★	★	★★★
Electromagnetic Steel Forming	★★★	★★★	★★★
Metal Spinning	★★	★ / ★★	★
Metal Cutting	★★	★★★	★
Industrial Origami®	★ / ★★ / ★★★	★ / ★★ / ★★★	★★★
Thermoforming	★ / ★★ / ★★★	★ / ★★ / ★★★	★★ (depends on mould)
Explosive Forming	★★	★ / ★★ / ★★★	★★
Superforming Aluminium	★★★	★★	★★★
Free Internal Pressure-Formed Steel	★	★★	★★
Inflating Metal	★	★★	★★★
Pulp Paper	★★★	★	★
Bending Plywood	★★ / ★★★	★ / ★★ / ★★★	★ / ★★ / ★★★ (depends on wood)
Deep Three-Dimensional Forming in Plywood	★★★	★★★	★
Pressing Plywood	★★	★	★★★

Type of shape	Size	Degree of tolerance	Relevant materials
Solid complex	S, M, L	★★★	Wood, metal, plastic. glass, ceramics
Solid complex. Any shape that can be produced on CAD	S, M, L	★★★	Virtually any material
Solid complex. Any shape that can be produced on CAD	S, M	★★★	Virtually any material (high melting temperatures slow down process)
Symmetrical	S, M	★★★ (metal) ★★ (other)	Ceramic, wood, metal, plastic
Solid	S, M	★	Ceramic
Sheet	S, M, L	★★	Electrically conductive metal
Sheet	S, M	★★★	Metal
Sheet	S, M	★★★	Plastic
Sheet	S, M, L	★★	Glass, metal, plastic, ceramic, stone, marble
Sheet	S, M, L	★★★	Conductive metal
Sheet	S, M	★★★	Metal, wood, plastic, paper, ceramic, glass
Sheet	S, M, L	★★	Ferrous metals, titanium
Sheet	S, M, L	★★	Metal
Sheet	S, M, L	★	Glass
Sheet		★★★	Magnetic metals
Sheet	S, M, L	★	Metal
Sheet	S, M, L	★★★	Metal
Sheet / Complex	S, M, L	★★★	Metal, plastic, composites
Sheet	S, M, L	★★	Thermoplastics
Complex	S, M, L	★★★	Metal
Sheet / Complex	S, M, L	★★	Superelastic aluminium
Hollow	M, L	★	Metal
Sheet	S, M, L	★★★	Metal, plastic
Sheet	M, L	★	Paper
Sheet	M, L	★	Wood
Sheet	S, M	★	Wood veneer
Sheet	S, M	N/A	Wood veneer

Key: ⋆= low ⋆⋆= mid ⋆⋆⋆= high	Cost of capital investment	Number of components produced per hour	Quality of surface finish
3. Continuous			
Calendering	⋆⋆⋆	⋆⋆⋆	⋆⋆⋆
Blown Film	⋆⋆⋆	⋆⋆⋆	⋆⋆⋆
Exjection®	⋆⋆⋆	⋆⋆⋆	⋆⋆⋆
Extrusion	⋆	⋆	⋆⋆⋆
Pultrusion	⋆⋆	⋆	⋆⋆
Pulshaping™	⋆⋆	⋆⋆⋆	⋆⋆
Roll Forming	⋆⋆⋆	⋆⋆⋆	⋆⋆⋆
Rotary Swaging	⋆	⋆⋆ / ⋆⋆⋆	⋆⋆⋆
Pre-Crimp Weaving	⋆	⋆ / ⋆⋆ / ⋆⋆⋆	⋆⋆
Veneer Cutting	N/A	N/A	⋆⋆
4. Thin & Hollow			
Glass Blowing by Hand	⋆ / ⋆⋆ / ⋆⋆⋆	⋆ / ⋆⋆	⋆⋆⋆
Lampworking Glass Tube	⋆	⋆ / ⋆⋆ / ⋆⋆⋆	⋆⋆⋆
Glass Blow and Blow Moulding	⋆⋆⋆	⋆⋆ / ⋆⋆⋆	⋆⋆⋆
Glass Press and Blow Moulding	⋆⋆⋆	⋆⋆ / ⋆⋆⋆	⋆⋆⋆
Plastic Blow Moulding	⋆⋆⋆	⋆⋆⋆	⋆⋆⋆ (parting lines remain)
Injection Blow Moulding	⋆⋆⋆	⋆⋆⋆	⋆⋆⋆
Extrusion Blow Moulding	⋆⋆⋆	⋆	⋆⋆⋆
Dip Moulding	⋆	⋆⋆ / ⋆⋆⋆	⋆
Rotational Moulding	⋆⋆	⋆⋆ / ⋆⋆⋆	⋆⋆
Slip Casting	⋆ / ⋆⋆ (depends on number of components)	⋆ / ⋆⋆ / ⋆⋆⋆	⋆⋆
Hydroforming Metal	⋆⋆⋆	⋆⋆⋆	⋆ / ⋆⋆ (depends on material)
Backward Impact Extrusion	⋆	⋆ / ⋆⋆	⋆⋆
Moulding Paper Pulp	⋆⋆⋆	⋆⋆⋆	⋆
Contact Moulding	⋆⋆⋆	⋆	⋆⋆ / ⋆⋆⋆ (depends on method)
Vacuum Infusion Process (VIP)	⋆⋆	⋆	⋆⋆⋆
Autoclave Moulding	⋆⋆	⋆⋆	⋆ / ⋆⋆ (if gel is applied)
Filament Winding	⋆	⋆ / ⋆⋆ / ⋆⋆⋆	⋆⋆ (finishing required)
Centrifugal Casting	⋆ / ⋆⋆ / ⋆⋆⋆ (depends on mould material)	⋆ / ⋆⋆	⋆ / ⋆⋆ (depends on process)
Electroforming	⋆	⋆	⋆⋆⋆

Type of shape	Size	Degree of tolerance	Relevant materials
Sheet	L	N/A	Textile, composite, plastic, paper
Sheet / Tube	L	★★★	LDPE, HDPE, PP
Continuous / Complex	S, M, L	★★★	Wood, plastic, metals
Sheet / Complex / Continuous	M, L	★★★	Plastic, wood-based plastic, composites, aluminium, copper, ceramic
Any shape of constant thickness	S, M, L	★★★	Any thermoset plastic combined with glass and carbon fibre
Variety of extruded cross sections, continuous	L	★★★	Thermosetting resins with glass, carbon or aramid fibre
Sheet	M, L	★★ / ★★★ (depends on thickness)	Metal, glass, plastic
Tube	S, M	★★	Ductile metals
Sheet	M, L	N/A	Any weavable alloy, mainly stainless or galvanized steel
Sheet / Continuous	M, L	N/A	Wood
Any	S	★	Glass
Symmetrical	S, M	★	Borosilicate glass
Simple forms	S, M	★	Glass
Simple forms	S, M	★★	Glass
Simple rounded forms	S, M, L	★★★	HDPE, PE, PET, VC
Simple forms	S	★★★	PC, PET, PE
Complex	S, M	★★	PP, PE, PET, PVC
Soft, flexible, simple forms	S, M	★	PVC, latex, polyurethanes, elastomers, silicones
Any	S, M, L	★	PE, ABS, PC, NA, PP, PS
Ranging from simple to complex	S, M	★	Ceramic
Tube, T-sections	S, M, L	★★★	Metal
Symmetrical	S, M	★★★	Metal
Complex	S, M, L	★★ / ★★★ (depends on process)	Paper: newspaper and cardboard
Open, thin cross-sections	S	★	Carbon, aramid, glass and natural fibres, thermosetting resin
Complex	M, L	★	Resin, fibreglass
Simple	S, M, L	★	Fibre and thermoset polymers
Hollow, symmetrical	L	★★★	Fibre and thermoset polymers
Tubular	S, M, L	★★★ (depends on process)	Metal, glass, plastic
Complex	S, M, L	★★★	Electroplatable alloys

Key: ⋆= low, ⋆⋆= mid, ⋆⋆⋆= high	Cost of capital investment	Number of components produced per hour	Quality of surface finish
5. Into Solid			
Sintering	⋆⋆/⋆⋆⋆	⋆⋆	⋆⋆⋆
Hot Isostatic Pressing (HIP)	⋆⋆	⋆	⋆⋆⋆
Cold Isostatic Pressing (CIP)	⋆⋆	⋆	⋆⋆
Compression Moulding	⋆⋆	⋆⋆⋆	⋆⋆⋆
Transfer Moulding	⋆⋆/⋆⋆⋆	⋆⋆⋆	⋆⋆⋆
Foam Moulding	⋆⋆⋆	⋆⋆⋆	⋆
Foam Moulding into Plywood Shell	⋆⋆	⋆⋆	N/A
Inflating Wood	⋆	⋆	N/A
Forging	⋆/⋆⋆/⋆⋆⋆	⋆/⋆⋆/⋆⋆⋆	⋆
Powder Forging	⋆⋆⋆	⋆⋆/⋆⋆⋆	⋆⋆
Precise-Cast Prototyping (pcPRO®)	⋆	⋆	⋆
6. Complex			
Injection Moulding	⋆⋆⋆	⋆⋆⋆	⋆⋆⋆
Reaction Injection Moulding (RIM)	⋆⋆⋆	⋆⋆	⋆⋆⋆
Gas-Assisted Injection Moulding	⋆⋆⋆	⋆⋆⋆	⋆⋆⋆
MuCell® Injection Moulding	⋆⋆⋆	⋆⋆⋆	⋆⋆⋆
Insert Moulding	⋆⋆⋆	⋆⋆⋆	⋆⋆⋆
Multi-Shot Injection Moulding	⋆⋆⋆	⋆⋆⋆	⋆⋆⋆
In-Mould Decoration	⋆⋆⋆	⋆⋆⋆	⋆⋆⋆
Over-Mould Decoration	⋆⋆⋆	⋆⋆⋆	⋆⋆⋆
Metal Injection Moulding (MIM)	⋆⋆⋆	⋆⋆	⋆⋆⋆
High-Pressure Die-Casting	⋆⋆⋆	⋆⋆⋆	⋆⋆⋆
Ceramic Injection Moulding	⋆⋆⋆	⋆⋆	⋆⋆⋆
Investment Casting	⋆⋆⋆	⋆⋆	⋆⋆⋆
Sand Casting	⋆	⋆⋆	⋆
Pressing Glass	⋆⋆	⋆⋆⋆	⋆⋆
Pressure-Assisted Slip Casting	⋆⋆⋆	⋆⋆⋆	⋆⋆⋆
Viscous Plastic Processing (VPP)	⋆⋆⋆	⋆⋆⋆	⋆⋆⋆
7. Advanced			
Inkjet Printing	⋆	⋆	⋆⋆
Paper-Based Rapid Prototyping	⋆	⋆	⋆
Contour Crafting	⋆	⋆	⋆
Stereolithography (SLA)	⋆	⋆	⋆
Electroforming for Micro-Moulds	⋆	⋆⋆⋆	⋆⋆⋆
Selective Laser Sintering (SLS)	⋆	⋆	⋆
Smart Mandrels™ for Filament Winding	⋆⋆	⋆	N/A
Incremental Sheet-Metal Forming	⋆	⋆⋆	⋆⋆⋆

Type of shape	Size	Degree of tolerance	Relevant materials
Complex / Solid	S, M	★★	Ceramic, glass, metal, plastic
Complex / Solid	S, M, L	★★	Ceramic, metal, plastic
Complex / Solid	S, M	★★	Ceramic, metal
Solid	S	★★	Ceramic, plastic
Complex / Solid	M, L	★★★	Composite, thermoset plastics
Complex / Solid	S, M, L	★★	Plastic
Solid	M, L	★★	Wood, plastic
Solid	M, L	★★	Wood, plastic
Solid	S, M, L	★	Metal
Complex / Solid	S, M, L	★★	Metal
Complex / Solid	S	★★★	Plastic
Complex	S, M	★★★	Plastic
Complex	S, M, L	★★★	Plastic
Complex / Solid	S, M, L	★★★	Plastic
Complex	S, M, L	★★★	Plastic
Complex	S, M	★★★	Plastic, metal, composite
Complex	S, M	★★★	Plastic
Complex	S, M	★★★	Plastic, metal, composite
Complex	S, M	★★★	Plastic, metal, composite
Complex	S, M	★★★	Metal
Complex	S, M, L	★★★	Metal
Complex	S, M	★★★	Ceramic
Complex	S, M, L	★★★	Metal
Complex	S, M, L	★	Metal
Hollow	S, M, L	★★★	Glass
Hollow	S, M, L	★★	Ceramic
Complex	S, M, L	★★	Ceramic
Sheet	S	★★★	Other
Complex	S	★★★	Other
Complex	L	★★★	Ceramic, composite
Complex	S, M	★★★	Plastic
Flat	S	★★★	Plastic
Complex	S, M	★★★	Metals, plastics
Hollow	S, M, L	★★	Plastic
Sheet	S, M, L	★	Metal

1:
Cut fro
Solid

The use of cutting tools to sculpt materials

This chapter encompasses some of the oldest processes used in the manufacture of objects, and these processes can be quite simply categorised by the fact that they use tools that cut away, shape and remove material. Increasingly, the 'brutal' part of these processes is being performed by automated CAD-driven machines, which carve effortlessly through most materials, providing yet another avenue for the exploitation of rapid prototyping technology and the replacement of the craftsman who gave life to many products throughout history.

Machining

including turning, boring, facing, drilling, reaming, milling and broaching

Machining belongs to a branch of production that falls under the commonly used umbrella term 'chip-forming' (meaning any cutting technique that produces 'chips' of material as a result of the cut). All machining processes have in common the fact that they involve cutting in one form or another. Machining is also used as a post-forming method, as a finishing method and for adding secondary details such as threads.

The term 'machining' itself embraces many different processes. These include several forms of lathe operation for cutting metals, such as turning, boring, facing and threading, all of which involve a cutter being brought to the surface of the rotating material. Turning (see also p.20) generally refers to cutting the outside surface, while boring refers to cutting an internal cavity. Facing uses the cutter to cut into the flat end of the rotating work piece. It is used to clean up the end face, but the same tool can be used to remove excess material.

Product	**Mini Maglite® torch**
Designer	Anthony Maglica
Materials	aluminium
Manufacturer	Maglite Instruments Inc.
Country	USA
Date	1979

The Maglite® torch, with its highly distinctive engineered aesthetic, has been produced using a number of metal chip-forming techniques, notably turning. The textured pattern for the grip, however, is produced post forming using a process known as knurling.

Volumes of production
These vary according to type, but computer numerical control (CNC)-automated milling and turning production involves several cutters working on several parts at the same time, which can result in reasonably high volumes of production. This large collection of techniques also includes hand machining of individual components.

Unit price vs capital investment
In general, there are no tooling costs involved, but the mounting and unmounting of work from the machine reduces production rates. However, the process can still be economical for short runs. CNC-automated milling and turning use CAD files to automate the process and produce complex shapes, which can be batched or mass-produced. Although standard cutters can be used for most jobs, specific cutters may need to be produced, which would drive up overall costs.

Speed
Varies depending on the specific process.

Surface
Machining involves polishing, to a degree, and it is possible to achieve excellent results without the need for post forming. Cutters can also produce engineered, ultra-flat surfaces.

Types/complexity of shape
Work produced on a lathe dictates that parts are axisymmetric, since the work piece is rotated around a fixed centre. Milled parts start life as a block of metal and allow for much more complex components to be formed.

Scale
Machined components range in size from watch components up to large-scale turbines.

Tolerances
Machined materials can deliver exceptionally high levels of tolerance: ±0.01 millimetre is normal.

Relevant materials
Machining is generally applied to metals, but plastics, glass, wood and even ceramics also make use of the machining process. In the case of ceramics, there are certain glass ceramics that are specifically designed to be machined and allow for new forms of processing ceramics. Macor is a particularly well-known brand. Mycalex, a glass-bonded mica by the US-based company Mykroy, is another machinable ceramic that eliminates the need for firing.

Typical products
Unique parts for industry – pistons, screws, turbines and a mass of other small and large parts for different industries. Alloy car wheels are often put on a lathe to finish the surface.

Similar methods
The term 'machining' encompasses such a wide set of processes that it is a family of methods in itself, but you could consider dynamic lathing (p.20) as an alternative to conventional lathing.

Sustainability issues
These processes are based only on mechanical energy and no heat so energy consumption is low. However, because the nature of these processes is the removal of material, a lot of waste is created. Depending on the material, waste can be reused or recycled.

Further information
www.pma.org
www.nims-skills.org
www.khake.com/page88.html

Threading is a process that uses a sharp, serrated tool to create screw threads in a pre-drilled hole.

Drilling and reaming are generally also lathe operations (though they can be also be done on a milling machine, or by hand), but they require different cutters. As with all lathe operations, the work piece is clamped in the centre of a rotating chuck. Whereas drilling is a straightforward operation to create a hole, reaming involves enlarging an existing hole to a smooth finish, which is done with a special reaming tool that has several cutting edges.

Other machining processes include milling and broaching. Milling involves a rotating cutter, similar to a drill, which is often used to cut into a metal surface (though it can be applied to just about any other solid material). Broaching is a process used to create holes, slots and other complex internal features (such as the internal shape of a spanner head, after it has been forged, see p.187).

1 A very simple set-up for milling a chunk of metal. The cutting tool, which resembles a flat drill bit, can be seen fitted above the clamped work piece.

2 A straightforward set-up for a lathe operation in which the tube of metal to be cut is clamped into a chuck. The cutter is poised ready to make a cut.

- Very versatile in terms of producing different shapes.
- Can be applied to virtually any solid material.
- High degree of accuracy.

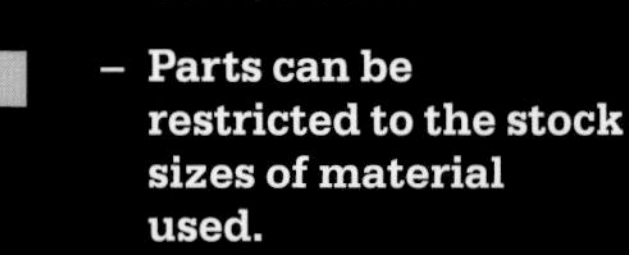

- Can be slow.
- Parts can be restricted to the stock sizes of material used.
- Low material utilisation due to wastage when cutting.

Computer Numerical Controlled (CNC) Cutting

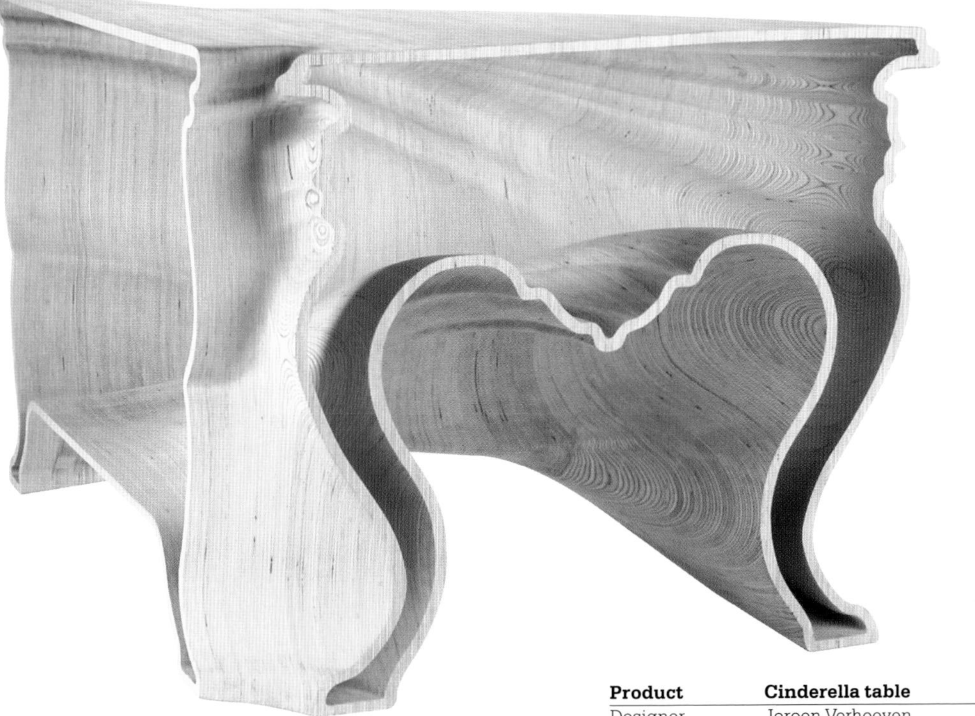

Product	Cinderella table
Designer	Jeroen Verhoeven
Materials	Finnish birch plywood
Manufacturer	Demakersvan
Country	Holland
Date	2004

The surreal construction and shape of this table from the 'Cinderella' range fits perfectly with the manufacturer's belief that high-tech machines are our hidden Cinderellas. The table is a witty play on traditional, romantic furniture made using a thoroughly modern manufacturing process.

The way computer numerical controlled (CNC) machines effortlessly cut through solid materials as if they were butter is almost sublime. The cutting heads are mounted onto a head that rotates in up to six axes, to chisel different forms as if they were automated robotic sculptors.

Designed by Jeroen Verhoeven, a member of the Dutch design group Demakersvan, the piece of furniture featured here is as multi-layered in meaning as it is in its construction.

As Demakersvan puts it, 'The big miracle of how industrial products come about is a wonderful phenomenon if you look at it closely. The high-tech machines are our hidden Cinderellas. We make them work in robot lines, while they can be so much more.'

This thought is put into practice in the production of its Cinderella table (pictured). The table is made up of 57 layers of birch multiplex, which are individually cut, glued and then cut again with a CNC machine. The table exemplifies perfectly the ability of multi-axis CNC machines to carve away at three-dimensional forms in a highly intricate manner, using information from a CAD file. It is also a unique example of a totally new form: created from an ancient material in a process that can cut virtually any shape from a piece of material, this table goes some way to reveal what Demakersvan describes as the 'secrets hidden in high-tech production techniques'.

1 The individual sheets of cut plywood are clamped together before being machined.

2 View showing the machined internal structure before the external surface is cut.

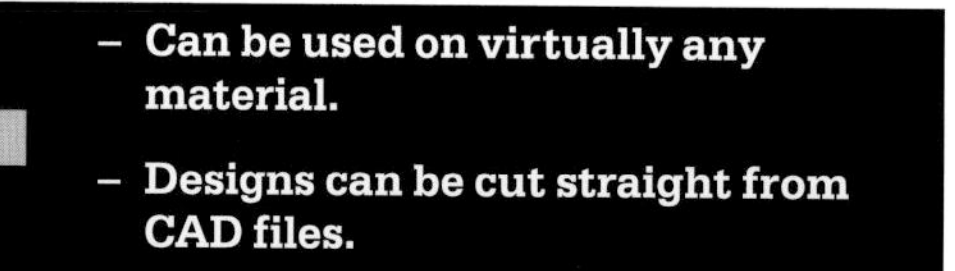

- Can be used on virtually any material.
- Designs can be cut straight from CAD files.
- Highly adaptable for cutting intricate and complex shapes.

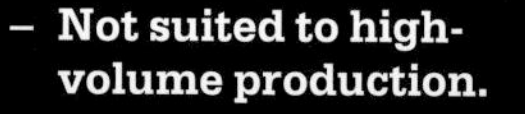

- Not suited to high-volume production.
- Can be slow.

Volumes of production
CNC cutting is best suited to one-off or batch production because of the slow progress rate.

Unit price vs capital investment
No tooling, just the expensive time for cutting and the creation, using CAD, of the three-dimensional data.

Speed
The speed is determined by several factors, including the material, the complexity of the form and the surface finish that is required.

Surface
Good, but may require some post finishing, depending on the material.

Types/complexity of shape
Virtually any shape that can be conceived on a computer screen.

Scale
From small components to huge objects. CNC Auto Motion in the US, for example, is one of several companies manufacturing monster-sized machines having over 15 metres of travel, with 3 metres of vertical-axis travel and a gantry measuring 6 metres across.

Tolerances
High.

Relevant materials
CNC technology can be used for cutting a wide range of materials, including wood, metal, plastic, granite and marble. It can also be used for cutting foam and modelling clay (see typical products).

Typical products
Ideally suited for making complex bespoke designs such as injection-moulding tools, die cutters, furniture components and the highly crafted, complex forms of handrails. It can also be used, in automotive design studios, for rapid prototyping of full-size cars in foam or modelling clay.

Similar methods
Laser cutting (p.46) when the laser is mounted on a multi-axis head is possibly the closest method.

Sustainability issues
As the machines have excellent accuracy the amount of wastage produced as a result of faulty parts or errors is minimal. Additionally, the layout of parts to be cut can be designed effectively so that there is minimal excess material. Depending on the material, waste can be reused or recycled.

Further information
www.demakersvan.com
www.haldeneuk.com
www.cncmotion.com
www.tarus.com

Electron-Beam Machining (EBM)

Product	customised tri-flange implants
Materials	titanium
Country	Sweden
Date	2005

This titanium hip-bone plate illustrates the spattered surface that is a common result of this process.

Electron-Beam Machining (EBM) is a versatile process that is used to cut, weld, drill or anneal components. As a machining process, one of its many advantages is that ultra-fine cuts can be made with such high precision

+

- **Highly accurate.**
- **There is no contact with the material being cut, so it therefore requires minimal clamping.**
- **Can be used for small batches.**
- **Versatile: a single tool can cut, weld and/or anneal at the same time.**

–

- **A disadvantage compared with laser cutting (see p.46) is that it requires a vacuum chamber.**
- **Laser cutting can be just as effective for less accurate machining.**
- **High energy consumption.**

that they can be measured in microns. EBM involves a high-energy beam of electrons being focused by a lens and fired at extremely high speeds (between 50 and 80 per cent of the speed of light) onto a specific area of the component, causing the material to heat up, melt and vaporise. The process needs to occur in a vacuum chamber to ensure that the electrons are not disrupted and thrown off course by air molecules.

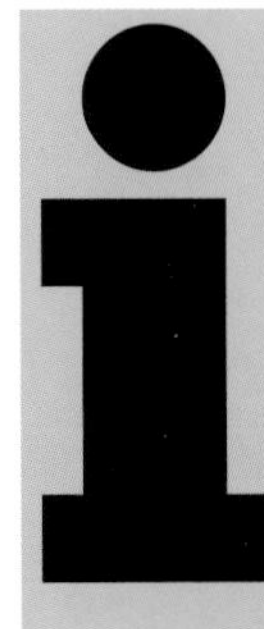

Volumes of production
Suited to one-off or batch production.

Unit price vs capital investment
Low capital investment as there are no tooling costs because the pattern is driven by a CAD file. However, the electron-beam equipment itself is very expensive.

Speed
The beam of electrons moves at a very high velocity, so cutting speeds are fast. For example, a hole of up to 125 microns in diameter can be cut almost instantly in a sheet 1.25 millimetres thick. Naturally, the type of material and its thickness affects the cycle time. In order to make a 100 millimetre-wide slot in a piece of stainless steel 0.175 millimetres thick, cutting will occur at a rate of 50 millimetres per minute. The implant (shown opposite) took four hours to produce.

Surface
The process can cause various surface markings that, depending on the application, might not be desirable, such as spattering close to the cut.

Types/complexity of shape
The process is ideally suited to cutting fine lines of holes in thin materials. The beam can be focused to 10 to 200 microns, which means that costs are justified by an extremely high degree of accuracy.

Scale
The disadvantage of using a vacuum chamber is that part sizes are limited.

Tolerances
Extremely high, with cuts as fine as 10 microns possible. With materials over 0.13 millimetres thick, the cut will have a fine, two-degree taper.

Relevant materials
Virtually any material, although materials that have high melting temperatures slow down the process.

Typical products
Apart from engineering applications and the medical implant shown here, one of the more interesting uses for EBM is for joining carbon nanotubes. Joining anything on the nano-scale is difficult, but because there is no contact with the material, EBM provides a method of joining the tubes together in a way that does not crush them.

Similar methods
Laser cutting (p.46) and plasma-arc cutting (p.33).

Sustainability issues
Very high amounts of energy are consumed to power the beam at such intensity and speeds. However, the versatility of electron-beam machining ensures that this energy is used effectively, as several processes can be carried out in one cycle. Additionally, as there is no contact with the material being cut, there is minimal damage or wear to the machine, which decreases material consumption through maintenance.

Further information
www.arcam.com
www.sodick.de

Turning
with dynamic lathing

The process of mounting a material on a spinning wheel and skimming off thin slices is thousands of years old. The commonly used material for turning is wood, but 'green' ceramic is also highly popular for industrially producing the same types of round, symmetrical shape.

In ceramic turning, a clay is blended into a ceramic body and extruded into something called a 'pug'. This leather-hard, clay lump is mounted onto a lathe and turned, either by hand or with an automated cutter.

At the other end of the industrial production scale, engineers at Germany's Fraunhofer Institute have developed a process called dynamic lathing for producing non-axisymmetric metal parts for engineering applications, without the need to remove and replace the component manually. Shapes are defined by a CAD program and fed directly to a lathe that allows the cutter to move up and down in the lateral axis.

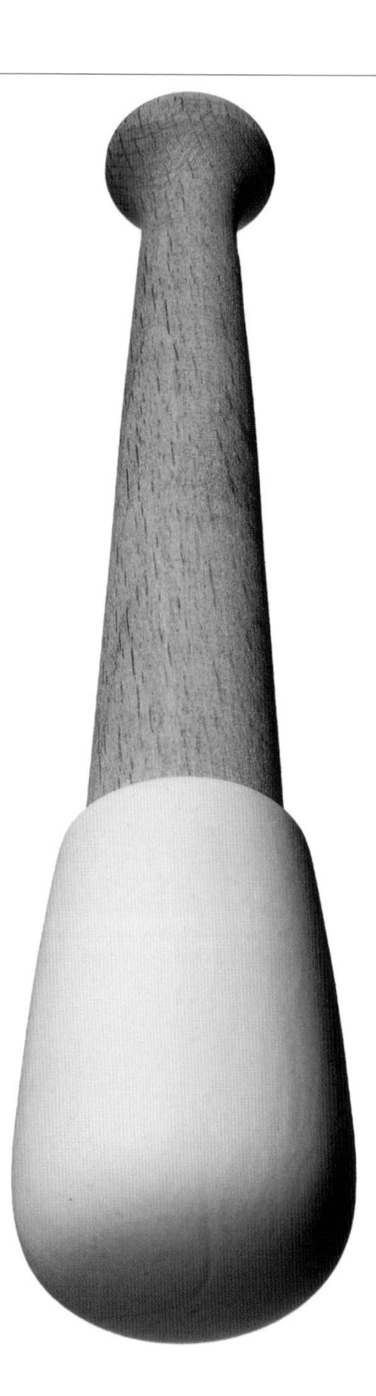

Product	**pestle**
Materials	ceramic stoneware, with wooden handle
Manufacturer	Wade Ceramics
Country	UK

The turning process has been used for both parts of this pestle, the wooden handle and the ceramic grinder head.

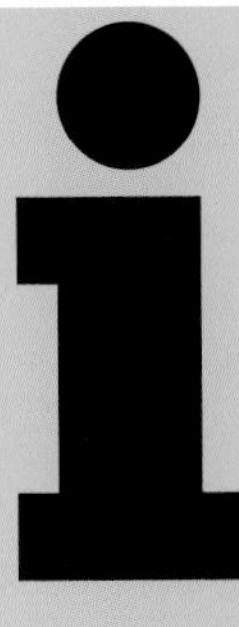

Volumes of production
From single pieces upwards. The costs of tooling and set-up for a single piece can be prohibitive, but this depends on specific requirements. For large quantities, an automated process may be required. The dynamic lathe is currently still in its infancy and is best suited to small production runs or one-offs.

Unit price vs capital investment
Depends on volume, but low in comparison to other ceramic production methods, such as hot or cold isostatic pressing (see pp.170 and 172) and slip casting (see p.140). In the dynamic lathe method there is no tooling, which obviously keeps costs down.

Speed
Depends on the product. As an example, a simple candlestick will take 45 seconds, a mortar 1 minute and a pestle 50 seconds. The relationship between the length and depth of the cut in the dynamic-lathe technique determines the speed at which parts can be made. The more peaks with larger depths there are, the slower the process is.

Surface
Fine surface, but dependent on the material (for example, the wooden handle is less fine than the ceramic head of the pestle shown here).

Types/complexity of shape
Restricted to symmetrical shapes. Dynamic lathes are a marked improvement on conventional turning on a metalwork lathe, and can produce far more complex parts than might traditionally have been made by casting.

Scale
A standard maximum size, as produced by Wade Ceramics in the UK, for example, is 350 millimetres in diameter by 600 millimetres in length. A maximum length of 300 millimetres with a maximum working diameter of 350 millimetres is possible with the dynamic-lathe method.

Tolerances
Tolerance ±2 per cent or 0.2 millimetres, whichever is the greater. However, this is higher when cutting metal on a lathe, especially if using CNC (computer numerical control). For dynamic lathe the tolerance is ±1 millimetre.

Relevant materials
Ceramics and wood are common materials for turning, however just about any solid material can be cut in this way. Most metals and plastics can be used for the dynamic-lathe process, although hard carbon steels can be problematic.

Typical products
Bowls, plates, door handles, pestles, ceramic electrical insulators and furniture.

Similar methods
For ceramics, a similar type of rotating set-up is used in both jiggering and jollying (p.29).

Sustainablity issues
The process is based on the removal of material to leave a three-dimensional form and therefore results in high quantities of wasted material. Depending on the material this may or may not be recyclable.

Further information
www.wade.co.uk
www.fraunhofer.de/fhg/EN/press/pi/2005/09/Mediendienst92005Thema3.jsp?print=true

1 The mortar bowl is being turned by hand, using a profiled metal tool to achieve a precise profile.

2 A ceramic pestle being finished using a flat smoothing tool.

+

- Low- or high-volume production runs.
- Can be used for a range of materials.
- Can have low tooling costs.
- Dynamic-lathe process allows non-round shapes to be cut in a single lathe operation.

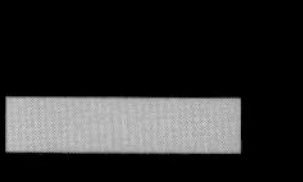

- In standard turning, parts are limited to circular profiles.
- In the dynamic-lathe process, the surface finish is compromised by the depth of cut and the number of peaks, both of which also contribute to a reduction in speed.

Jiggering and Jollying

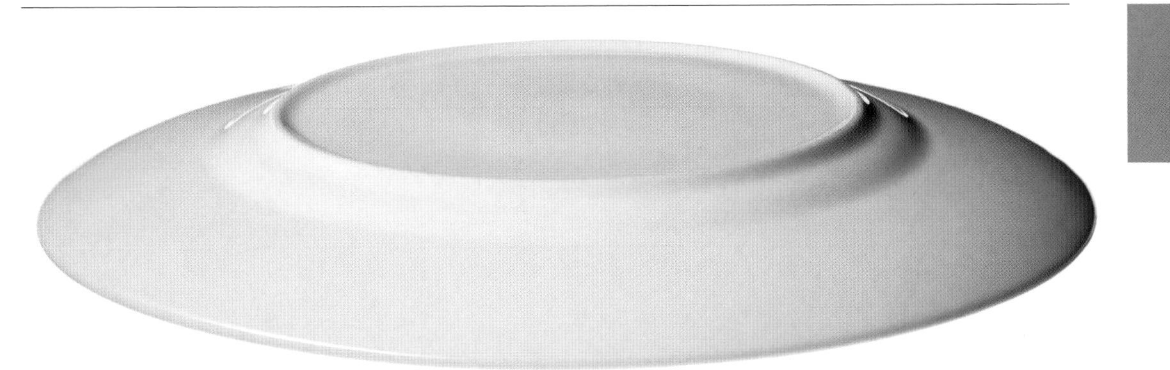

Jiggering and jollying are two profoundly silly words that describe similar methods for the mass-production of ceramic hollow shapes, such as bowls, or flatter shapes, such as plates. The easiest way to get a sense of these methods is to think of hand throwing on a potter's wheel, but turned into an industrial process where the craftsman's hands are replaced by a profiled cutter, which scrapes the clay as it rotates on the wheel. In jiggering, the mould determines the internal form of the shape while the cutter forms the outer shape, while in jollying the cutter forms the inner shape.

Jollying is employed to make deep shapes, the first stage of which involves extruding a clay slug that is cut into discs and used to form liners. These are like clay cups that are formed to be of similar proportions to the final piece. The liners are placed inside the cup moulds, which are fitted to a rotating spindle on the jollying wheel. This is where the similarity to hand throwing comes in. On the rotating spindle, the clay is drawn up the inside of the mould, forming the wall. A profiled head is then brought down into the cup to scrape away the clay and form the finished and precise inside profile.

Jiggering is a very similar process to jollying but is used to form shallow rather than deep shapes. It works in an inside-out way to jollying, because the shaped profile cuts the outside surface rather than the inside. Again, a slug of clay is formed and placed over a rotating mould, known as

Product	Wedgwood® plate
Materials	bone china
Manufacturer	Wedgwood®
Country	UK
Date	1920

This classic design from the Wedgwood® stable is produced using jiggering, a process that has altered little since the foundation of Josiah Wedgwood's pottery in 1759, except for the introduction of electricity to power the wheel. Turn any such plate upside down and you see the shape of the cutting profile used to scrape away the clay.

a 'spreader'. Here, it is formed into an even thickness by a flat profile. This thick pancake, which is known as a 'bat', is removed and placed onto the plate mould. The mould forms the inside shape of the plate. The whole thing rotates and a profile is brought down to scrape away the external side of the clay and form a precise, uniform outside shape.

Volumes of production
Can be used for both batched and mass-production. Many of the big potteries use these methods as a standard way of producing bowls and plates.

Unit price vs capital investment
Affordable tooling for batch production. The process can also be used for small runs of handmade production.

Speed
Jollying produces an average of eight pieces per minute, jiggering an average of four units per minute.

Surface
The surface finish is such that the products can be glazed and fired without any intermediate finishing.

Types/complexity of shape
Compared with slip casting (see p.140), where the detail on the inside wall of a piece is totally dependent on the outside form, these two processes allow total control over both the inside and outside profiles individually.

Scale
The standard size for machine-made dinner plates is up to a fired diameter of approximately 30 centimetres.

Tolerances
± 2 millimetres.

Relevant materials
All types of ceramic.

Typical products
Both methods are principally used for producing tableware and are distinguished by the depth of the shapes they produce. Jollying is used to make products such as pots, cups and bowls, which are generally deep containers, while jiggering is used to make shallow items such as plates, saucers and shallow bowls.

Similar methods
Apart from using a potter's wheel, the closest ceramic alternative to jiggering and jollying is turning (p.26), which can be used to make symmetrical shapes and different profiles without a large investment in tooling (although it requires a more complex set-up). Alternative methods also include cold and hot isostatic pressing (pp.170 and 172) and pressure-assisted slip casting (p.140).

Sustainability issues
The excess clay that is removed from the mould can be reused, which reduces overall material consumption. Firing the clay at high temperatures is very energy intensive, while the initial forming is not. However, several hundred pieces can be fired in a single kiln to make the most efficient use of energy.

Further information
www.wades.co.uk
www.royaldoulton.com
www.wedgwood.co.uk

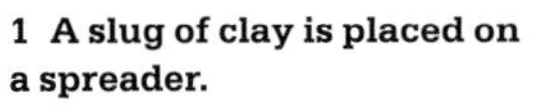

1 A slug of clay is placed on a spreader.

2 A flat profile forms it into an even 'bat'.

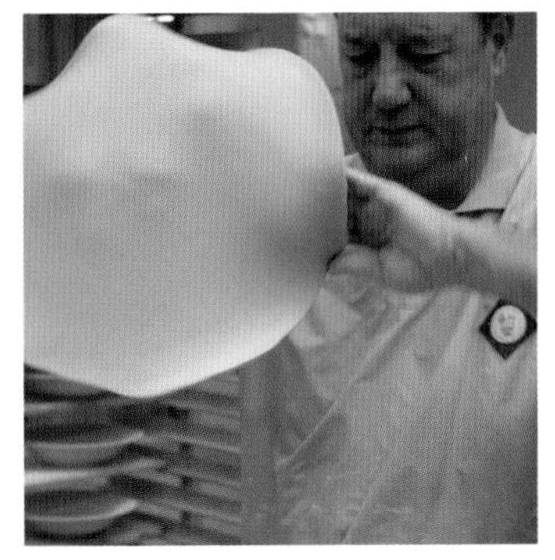

3 The bat is transferred by hand from the spreader.

4 Manual placement of the bat onto the mould.

5 The profiled head, closely supervised, acts against the rotating bat to scrape away the clay.

6 Typical of the sort of flat shape produced by jiggering, these soup bowls are now ready for firing.

1 A rough disc (liner) is pressed into a deep mould.

2 A profiled head spreads the liner evenly around the inside of the rotating mould.

3 Hand finishing of the outer surface, the side that was in contact with the mould.

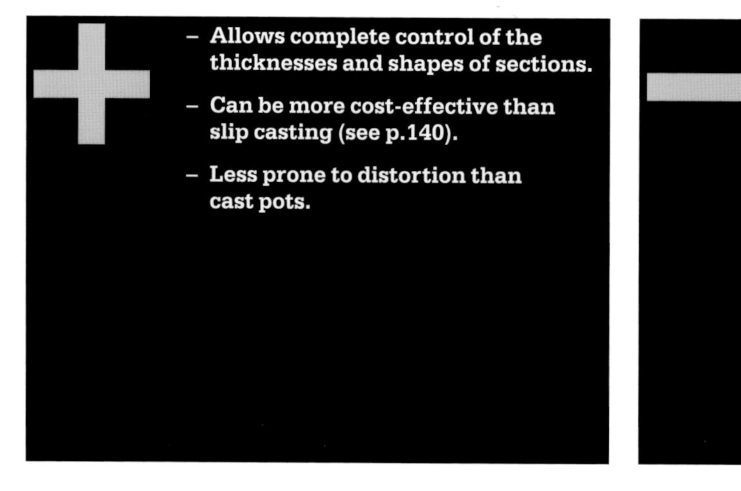

- Allows complete control of the thicknesses and shapes of sections.
- Can be more cost-effective than slip casting (see p.140).
- Less prone to distortion than cast pots.

- Can be inaccurate due to shrinkage during firing.
- Because these processes both work on the potter's-wheel principle, they are only able to produce symmetrical parts.

Plasma-Arc Cutting

'Men in overalls wearing dark view-control helmets' is perhaps all I need to say to sum up this process. Along with oxyacetylene cutting (see p.48), plasma-arc cutting lives in the land of heavy industry, and it is part of the non-chip-forming branch of production known as thermal cutting. It works by means of a stream of ionised gas, which becomes so hot that it will literally vaporise the metal that is being cut.

The process takes its name from the term 'plasma', which is what a gas turns into when it is heated to a very high level. It involves a stream of gas – usually nitrogen, argon, or oxygen – being sent through a small channel at the centre of a nozzle, which at its heart contains a negatively charged electrode. The combination of power supplied to this electrode, and contact between the tip of the nozzle and the metal being cut, results in a circuit being created. This produces a powerful spark, the arc, between the electrode and the metal work piece, which heats the gas to its plasma state. The arc can reach temperatures as high as 27,800°C and it therefore melts the metal as the nozzle passes over it.

This shows the heavy, industrial nature of this particular cutting method. The tube is rotated around the central axis to allow for a short length to be removed.

The cutting line width, known as the 'kerf', needs to be considered when designing certain shapes because its thickness can range from 1 to 4 millimetres, depending on the thickness of the metal plate, and can affect the dimensions of the component.

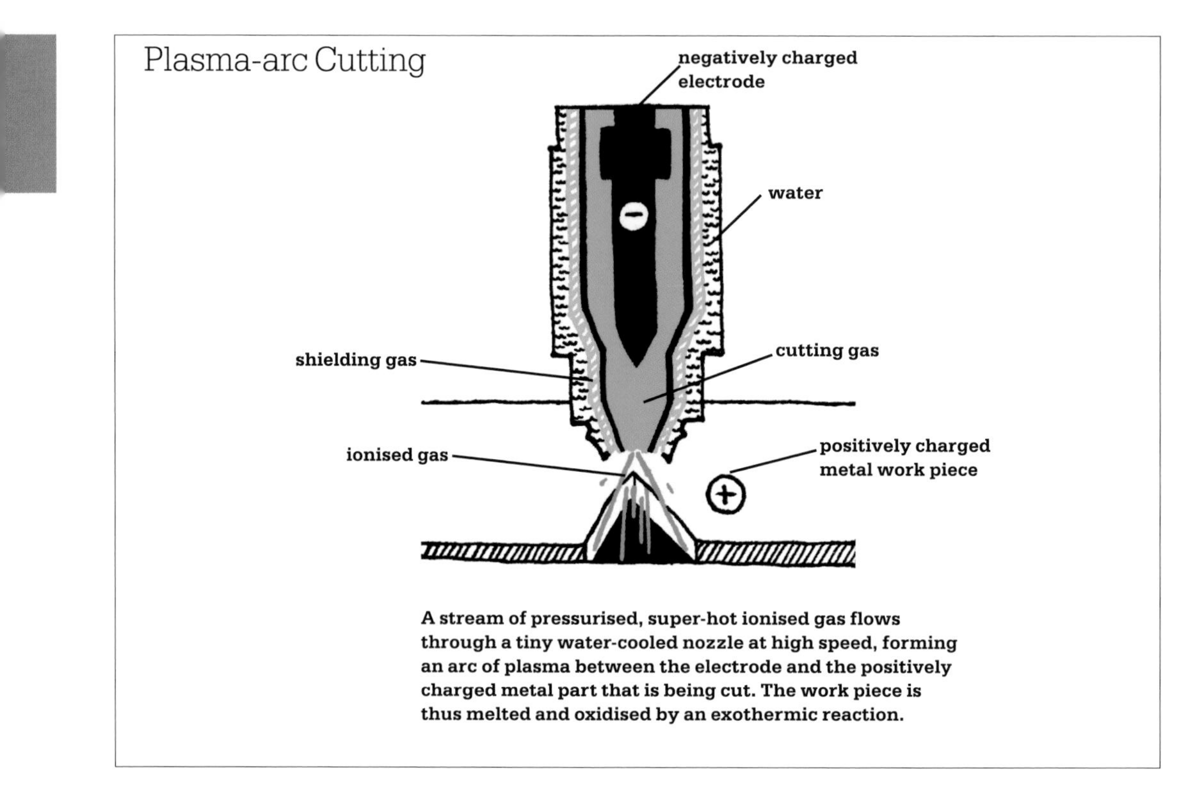

A stream of pressurised, super-hot ionised gas flows through a tiny water-cooled nozzle at high speed, forming an arc of plasma between the electrode and the positively charged metal part that is being cut. The work piece is thus melted and oxidised by an exothermic reaction.

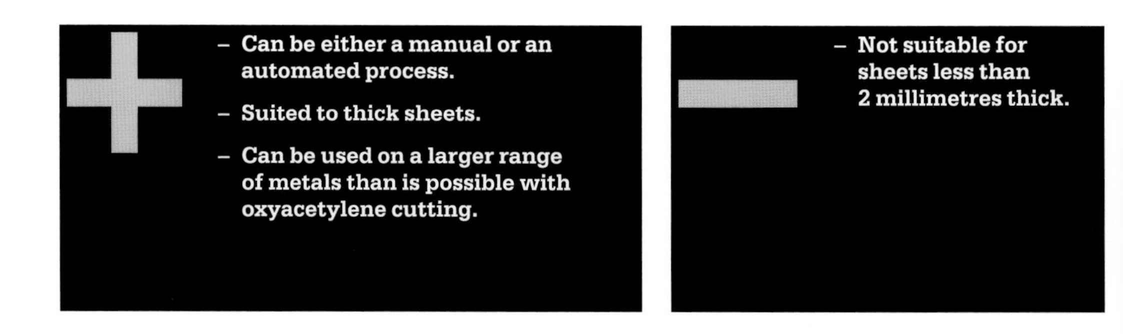

- Can be either a manual or an automated process.
- Suited to thick sheets.
- Can be used on a larger range of metals than is possible with oxyacetylene cutting.

- Not suitable for sheets less than 2 millimetres thick.

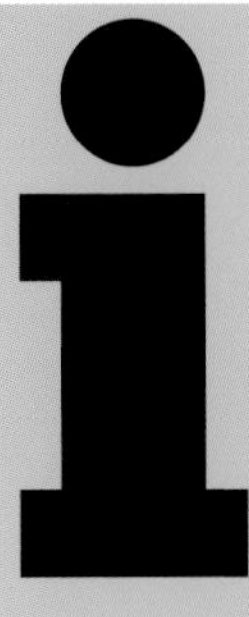

Volumes of production
Plasma-arc cutting is an economical process for small-batch quantities because it can be performed without tooling.

Unit price vs capital investment
Unless a cutting template is introduced, the process does not require tooling.
In the automated process, the information for the shape is provided by CAD files.

Speed
There is generally very little set-up time involved, but the speed is greatly affected by the type of material and its thickness.
For example, to cut a 25 millimetre chunk of steel, 300 millimetres long, will take about 1 minute, whereas a 2 millimetre piece can be cut at a rate of 2,400 millimetres per minute.

Surface
Even when hard stainless steel is used, the process provides smooth, clean edges with better results than those produced by oxyacetylene cutting (see p.48). The cutting can also be controlled to produce different grades of surface, depending on the cost versus the edge quality – that is, longer cutting times equal better edge finish.

Types/complexity of shape
The process is best suited to heavy-gauge materials. Thin-gauge metals of below 8 millimetres may distort as a result of the process, as may thin and narrow sections.
As in all sheet-cutting operations, nesting one shape within another (as when you make biscuits and cut them as close as possible to make the most of the rolled dough) results in an economical use of the material.

Scale
Handheld cutting means there is no maximum size. Sheets thinner than about 8 millimetres may distort.

Tolerances
Depend on the thickness of the material, but, to give a basic idea, it is possible to keep tolerance to ±1.5 millimetres for sheet materials 6–35 millimetres thick.

Relevant materials
Any electrically conductive metallic material, but most commonly stainless steel and aluminium. The process becomes more difficult the higher the carbon content of the steel.

Typical products
Heavy construction, including shipbuilding and machine components.

Similar methods
Electron-beam machining (EBM) (p.24), oxyacetylene cutting (p.48), laser (p.46) and water-jet cutting (p.42).

Sustainability issues
One of the most energy-intensive processes around, plasma-arc cutting requires extreme heat and pressure for the gas to achieve cutting strength. Because shapes can sometimes be cut from sheets there is also a high amount of material wastage.

Further information
www.aws.org
www.twi.org.uk/j32k/index.xtp
www.iiw-iis.org
www.hypertherm.com
www.centricut.com

2: Sheet

Components that start life as a sheet of material

Within the last fifteen years or so, there has been a surge in the number of products made from sheet material. Maybe this is because the starting point is a pre-prepared material, which goes some way in reducing the production costs. Maybe it is also the cost-effectiveness of die-cutting tools, or even the absence of any tooling costs for processes such as chemical milling. But, on a mass-market level, the die cutting of a plastic such as polypropylene has led to a wealth of new packaging, lighting and even larger-scale furniture. Perhaps it is also the ability of these materials to be cut by a manufacturer and handfolded and assembled by the consumer that has created an appeal.

Chemical Milling
AKA Photo-Etching

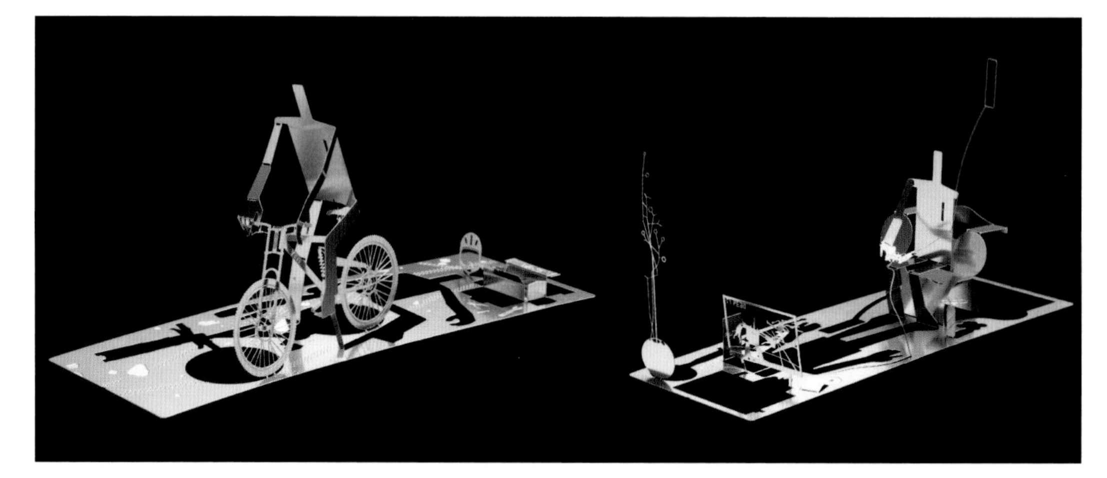

Product	Mikroman business card
Designer	Sam Buxton
Materials	stainless steel
Date	2003

The fine, intricate details of these cleverly designed business cards, which unfold to show a man on a bicycle and one in an office environment, are an excellent example of the ability of chemical milling to offer a highly decorative and ultra-fine method of cutting metals.

Chemical milling, also known as photo-etching, is a great method for producing intricate patterns on thin, flat metal sheets by using corrosive acids in a process similar to that used for developing photographs.

Chemical milling involves a resist being printed onto the surface of the material to be treated. This resist works by providing a protective layer against the corrosive action of the acid, and it can be applied in the form of a linear pattern or a photographic image. When

- No tooling.
- Highly flexible process for creating surface detail.
- The image is laser-plotted onto film from a CAD file, so designs can easily be modified.
- Fine tolerances.
- Suitable for thin sheets.

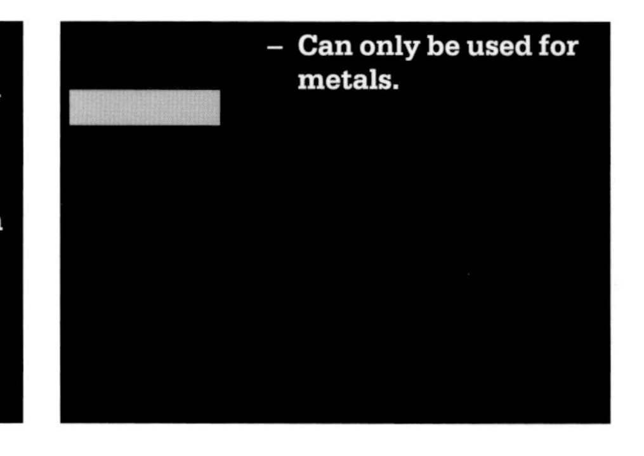

- Can only be used for metals.

the part is sprayed with an acid on both sides, the exposed metal (without the resist) is eaten away by the chemical. As in the die-cutting process (see p.40) used for plastics, crease lines can be 'half-etched' into the pattern, which allow the sheet to have foldable crease lines for the creation of three-dimensional structures.

Volumes of production
Individual pieces can be made, but the process is more suited to batch or mass-production volumes.

Unit price vs capital investment
The set-up costs involved are low, because the printed resist removes the need for any hard tooling. However, the unit costs are not likely to be drastically lower for batch items than for mass-produced items.

Speed
Depends on the complexity of the artwork.

Surface
Due to the corrosion of the metal, any half-etched surface has a rough and matt texture. However, this texture often becomes a decorative feature. Cut edges are free of burring.

Types/complexity of shape
The process is ideal for cutting thin sheets and foils. It also allows for highly intricate shapes and details to be cut without any blemishes such as the burn-marks that are sometimes caused by laser cutting (see p.46).

Scale
Generally limited to standard sheet sizes.

Tolerances
Tolerance levels are determined by the thickness of the material. The holes must be larger (typically 1 to 2 times) than the thickness of the metal, which gives a material with a thickness of between 0.025 and 0.050 millimetres a tolerance of ±0.025 millimetres.

Relevant materials
A range of metals can be used in the process, including titanium, tungsten and steels.

Typical products
Electronic components such as switch contacts, actuators, micro screens and graphics for industrial labelling and signs. The process is also used for industrial components, and the military use it to make a flexible trigger device for missiles. The trigger is so fine that it changes according to air pressure the closer it gets to its target.

Similar methods
Electron-beam machining (p.24), laser cutting (p.46), blanking (see metal cutting, p.59) and electroforming for micro-moulds (p.250).

Sustainability issues
Although the process involves the use of harmful chemicals, responsible disposal has been introduced through a cleaning process that removes contaminants and allows the liquids to be recycled back into the process. This reduces waste along with water consumption. Additionally, the complexity and accuracy of the etching means no secondary processing is necessary, saving further resources and energy. Unlike with mechanical cutting methods waste material is not available to recycle.

Further information
www.rimexmetals.com
www.tech-etch.com
www.precisionmicro.com
www.photofab.co.uk

Die Cutting

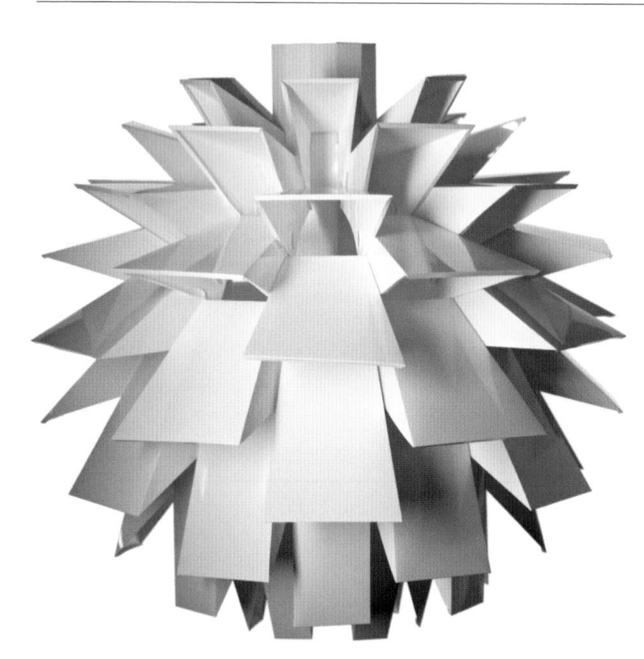

The simplest analogy for this process is to think of a biscuit-cutter for making shapes from dough in the kitchen. Just as easily applied to paper or plastic, die cutting is a simple process that involves a sharp edge being brought down onto a thin material to cut a shape in a single step. A die-cutting tool has two functions: the main function is to cut a shape from the sheet; the second is to apply creases to the material to allow it to form an accurate bend. The creases are necessary when constructing three-dimensional shapes and integrated hinges from a sheet.

Product	**Norm 69 lampshade**
Designer	Simon Karkov
Materials	polypropylene
Manufacturer	Normann Copenhagen
Country	Denmark
Date	2002

The 69 lampshade (above) is sold, flat, in boxes of pizza-size proportions. The flat pieces of die-cut plastic that are contained in the box take the customer about 40 minutes to fold and assemble into this complex structure.

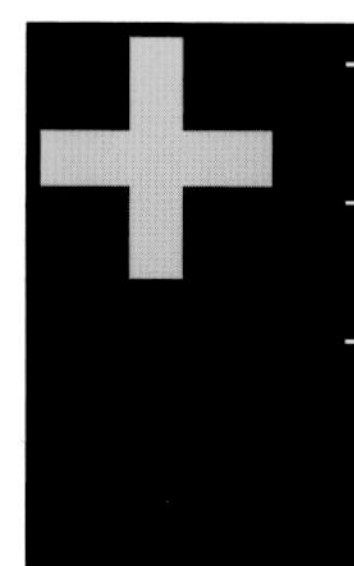

- **Low set-up costs and cost-effective for batches.**
- **Can easily be combined with printing.**
- **Many shapes can be cut in a single cutting action.**

–

- **Three-dimensional products need hand assembly and are limited to a set of standard constructions.**

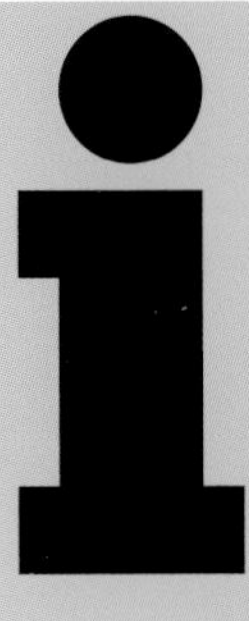

Volumes of production
From small batches of around a hundred units, to thousands.

Unit price vs capital investment
The low cost of the cutters makes this a highly economical process even for small runs. Sheets of material may be fed individually but if the material comes on a roll there will be a massive reduction in the cost of the final products.

Speed
Die cutting is one of the predominant manufacturing processes for packaging, with production cycle times of up to thousands of products per hour. Unlike in moulded products, the cutting speed is not affected by the complexity of the shape. Assembly, however, is more labour-intensive.

Surface
The surface is dependent on the material. The cut edge, however, is clean, precise and with a very, very fine radius where the cutter has cut through the material. As you might expect, the sheets can be finished with various forms of printing or embossing, or a combination of the two.

Types/complexity of shape
The complexity of the shape is really dependent on the size of the cuts. Very fine slots of less than about 5 millimetres can be difficult to cut. One of the design issues to bear in mind is that the excess plastic around the part needs to be removed and cleaning plastic from fine holes can be difficult.

Scale
Most manufacturers should have no problems at all in cutting sheets of up to 1,000 by 700 millimetres, and some are able to go slightly larger and cut straight from a roll. However, material choice is more limited if it comes on a roll. Printing on large sheets, above 1,000 by 700 millimetres, may be difficult due to the limited availability of large-scale printing machines.

Tolerances
Very high tolerances.

Relevant materials
A large proportion of the material used is polypropylene due to its ability to form a strong integral hinge. Other standard materials include PVC, polyethylene terephthalate (PET), paper and all sorts of card.

Typical products
Die cutting is extensively used for packaging, especially boxes and cartons. For this type of product, assembly is required to construct the three-dimensional structures. Other, more product-focused, applications include lampshades that require complex assembly (pictured), toys and even furniture.

Similar methods
For cutting flat sheets, try laser (p.46) or water-jet cutting (p.42).

Sustainability issues
Die cutting can be more economical in terms of material use if shapes are nested and material wastage is reduced. The nature of the material determines whether waste (of which there will be a considerable amount) can be reheated and recycled.

Further information
www.burallplastec.com
www.ambroplastics.com
www.bpf.co.uk

Water-Jet Cutting AKA Hydrodynamic Machining

From as early as the mid-nineteenth century, water jets have been used as a method of removing materials during mining. The modern-day process (also known as hydrodynamic machining) has been cranked up to produce an incredibly fine jet of water, typically 0.5 millimetres, which is forced out of a nozzle at a pressure of 20,000–55,000 psi (pounds per square inch) at velocities of up to twice the speed of sound. Water-jet cutting produces a fine cut using water alone, but with an additional abrasive, such as garnet, it can be used to cut through harder materials.

Product	**Prince chair**
Designer	Louise Campbell
Materials	water-cut EPDM (ethylene propylene diene monomer) on laser-cut metal, and felt
Manufacturer	Hay
Country	Denmark
Date	2005

The decorative pattern on this chair illustrates the potential of water-jet cutting to cut intricate patterns into a three-dimensional material.

+

- A cold process, so it does not heat the material.
- No tool contact, therefore no edge deformation.
- Can be used to cut very fine details in a variety of materials of different thicknesses.

−

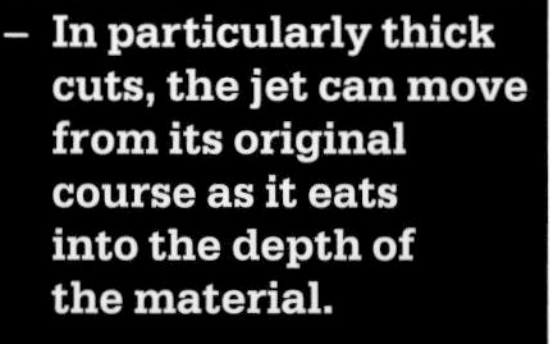

- In particularly thick cuts, the jet can move from its original course as it eats into the depth of the material.

Volumes of production
The process involves no tooling and is therefore equally suitable for one-off jobs and large production runs.

Unit price vs capital investment
Because there is no tooling, and designs are taken from CAD files, set-up costs are low, and there is no consequent hike in unit price. Shapes can also be 'nested' – laid out in a clever way that maximises the surface area of the sheet (as you would when cutting out biscuits from rolled dough).

Speed
An abrasive jet can cut a 13 millimetre-thick titanium sheet at the rate of 160 millimetres per minute.

Surface
The cut edge has the same sort of edge as if it had been sand-blasted, but without any of the burring that you may get with laser cutting (see p.46).

Types/complexity of shape
Because the cutter works like a plotting machine or a CNC router, it is possible to cut fine and intricate shapes. However, due to the high pressure of the water, thin sheet material may distort or bend. Processes such as laser cutting avoid this problem due to the lack of such pressure.

Scale
Most industrial cutting takes place on a cutting bed, which restricts the size of the material that can be used. Standard sizes extend up to a maximum of 3 by 3 metres. The upper limit for thickness varies with the material.

Tolerances
The jet can be accurate to 0.1 millimetre. Particularly thick materials may result in the jet 'wandering' slightly from its point of entry.

Relevant materials
Water-jet cutting offers a huge range of possibilities in terms of materials – you can choose from glass, steel, wood, plastic, ceramics, stone, marble and even paper. It is also used for cutting sandwiches and other food. Having said that, it is worth bearing in mind that materials that are particularly prone to absorbing water are not suited to this process.

Typical products
Decorative architectural panels and stones. The process works very well under water, and it was used in the rescue operation of the Russian Kursk submarine in 2000.

Similar methods
The process can be used as an alternative to die cutting (p.40), and as a cold alternative to laser cutting (p.46).

Sustainability issues
The water used for cutting can be recycled back into the process to form a closed loop cycle, which reduces water consumption and resources. Additionally, with no tool contact, maintenance and material use through replacement parts is reduced. No heat is required so energy use is fairly low while no fumes, toxins or contaminants are released during cutting. The nature of the material determines whether waste (of which there will be a considerable amount) can be reheated and recycled.

Further information
www.wjta.org
www.tmcwaterjet.co.uk
www.waterjets.org
www.hay.dk

Wire EDM (Electrical Discharge Machining) and Cutting with Ram EDM

Designers are rediscovering the use of surface decoration as a valid form of design expression. Industrial techniques have been borrowed from engineering applications and are used to create highly intricate patterns, as decorative as if they have been abstracted from nature or fairy-tale narratives.

Many unusual methods have been invented for cutting complex patterns into difficult materials. Dating back to when the phenomenon was first observed in the 1770s, electricity has been harnessed by scientists for use in cutting and machining materials. Wire EDM (electrical discharge machining) is one of the latest processes to exploit electricity for cutting intricate patterns.

Since its commercial development in the 1970s, wire EDM has become an increasingly popular method of machining metals. Together with processes such as water-jet cutting (see p.42) and laser cutting (see p.46), Wire EDM is a non-contact method of cutting materials. Less commonly used than the other two and more suited to extremely hard steels and other hard-to-cut metals such as high-performance alloys, carbides and titanium, it is, nevertheless, able to achieve the same level of intricacy.

Based on a type of spark-erosion (it is sometimes also referred to as spark machining or spark eroding), wire EDM is used to cut very hard, conductive metals by using sparks to melt away the material. The spark is generated by a thin wire – the electrode – which follows a programmed cutting path (determined by a CAD file). There is no contact between the electrode and the material, so the spark jumps across the gap and melts the material. De-ionised water is simultaneously jetted towards the melting point, cooling the material and washing the waste away.

There is another sort of EDM machine, the 'ram'. As the name suggests, the ram method involves a machined graphite electrode mounted on the end of an arm (the ram) being pushed onto the surface of the material to be cut.

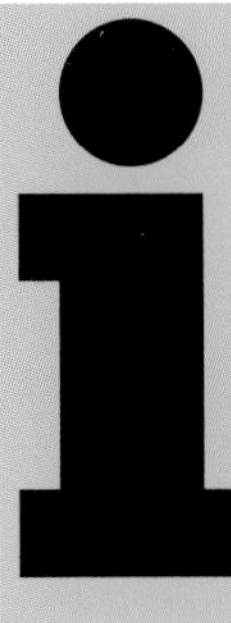

Volumes of production
The process and shape can be controlled manually by an operator or from a CAD file, so it is equally suited to one-off pieces and automated mass-production.

Unit price vs capital investment
Requires no tooling.

Speed
The latest generation of EDM machines can cut up to 400 square millimetres per minute, depending on the electrical resistance of the material and, of course, its thickness. A 50-millimetre piece of steel can be cut at a rate of approximately 4 millimetres per minute.

Surface
Wire EDM is well known for its ability to achieve an excellent finish.

Types/complexity of shape
The delicate wire can cut very intricate shapes from the toughest materials.

Scale
Depending on the material, the generator size and the power, the process can cut through massive hunks of metal up to an astonishing 500 millimetres thick, although this will be very time-consuming, with cutting occurring at a rate of less than 1 millimetre per minute.

Tolerances
Wire EDM is extremely accurate and can achieve sub-micron tolerances.

Relevant materials
Restricted to conductive metals. The process is ideally suited to hard metals, the hardness of which does not affect the cutting speed.

Typical products
One of the big markets for this process is for the super-hardened dies and cutters that are used in industrial production. Other applications include super-tough components for the aerospace industry.

Similar methods
Laser cutting (p.46) and electron-beam machining (EBM) (p.24).

Sustainability issues
Power consumption is very high, especially as the cutting rate is slow, which results in long cycle times. The offcut materials need to be melted down and recycled back into the process to reduce material consumption and waste.

Further information
www.precision2000.co.uk
www.sodick.com
www.edmmachining.com

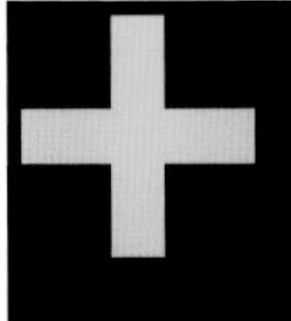

- **Ideal for cutting intricate shapes from metals that would be difficult to machine.**
- **The process cuts without force.**
- **No flushing.**

- **Time-consuming.**
- **Limited to electrically conductive materials.**

Laser Cutting
with laser-beam machining

Similar to water-jet cutting (see p.42) and electron-beam machining (see p.24), laser cutting is a non-chip-forming method of cutting and decorating materials. It is a highly accurate process based on input from a CAD file. In a nutshell, it works through a highly focused beam of light generating millions of watts of energy per square centimetre, which melts the material that is in its path.

Laser-beam machining is a form of laser cutting that uses a multi-axis head to cut three-dimensional objects. A CAD file maps complex paths for the powerful beam of light, resulting in fine, accurate designs.

Both of these processes are capable of cutting components that could not be cut precisely with conventional machine tools. As neither method involves contact with the material being cut, minimal clamping is required.

Product	**Spiral**
Designer	Torafu Architects
Materials	paper
Manufacturer	Kamino Kousakujo
Country	Japan
Date	2010

This is a paper bowl that envelops air. You can freely change its shape by moulding it. The thin and lightweight paper gains tension and strength when pulled out. The soft white colour and delicate expressions of this airvase allow it to subtly blend into any scene.

+

- **No tool wear, minimal clamping and it offers a consistent, highly accurate cut.**
- **Suited to a range of materials.**
- **No post treatment of edges.**

–

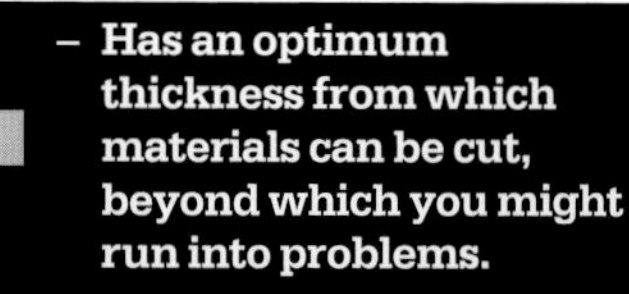

- **Has an optimum thickness from which materials can be cut, beyond which you might run into problems.**
- **Can be time-consuming on large production runs, so it is best suited to one-off or batch production.**

Volumes of production
Suited to batch production.

Unit price vs capital investment
Low capital investment as there is no tooling because the cuts are determined by a CAD file.

Speed
As with all methods of cutting, the speed of this process is dependent on the type of material used and its thickness. As a rough estimate, titanium alloys between 0.5 and 10 millimetres thick can be cut at a rate of 2.5 to 12 metres per minute.

Surface
The process will leave burn marks on wood, but on metal can give a clean edge with no need for post finishing. However, metal surfaces should be left unpolished before cutting, as highly polished surfaces act as reflectors and decrease the effectiveness of the process.

Types/complexity of shape
Depending on the machinery, the laser can be mounted horizontally or on a multi-axis head, allowing for highly complex shapes to be cut in three dimensions, a method that is sometimes called laser-beam machining.

Scale
Limited to standard sheet sizes.

Tolerances
Tolerances are extremely high, with holes of as little as 0.025 millimetre in diameter being possible.

Relevant materials
Often used on hard steels such as stainless and carbon steel. Copper, aluminium, gold and silver are more difficult due to their ability to conduct the heat. Non-metallics can also be laser cut, including woods, paper, plastics and ceramics. Materials such as glass and ceramics are especially suited to laser cutting, since it would be difficult to cut the materials in intricate patterns using any other techniques.

Typical products
Model components, surgical instruments, wooden toys, metal meshes and filters. Laser-cut ceramics can be used as industrial insulators and furniture can be produced using laser-cut glass or metal.

Similar methods
Water-jet cutting (p.42), die cutting (p.40), electron-beam machining (EBM) (p.24) and plasma-arc cutting (p.33).

Sustainability issues
Laser cutting is very energy intensive in order to sustain the beam intensity, and the speed is considerably slower when working with thick or large pieces. However, as no contact is made between the tool and the substrate, maintenance is low and this reduces material consumption through replacement parts. As with all sheet cutting techniques, material wastage is often high. The nature of the material determines whether waste can be reheated and recycled.

Further information
www.miwl.org.uk
www.ailu.org.uk
www.precisionmicro.com

Oxyacetylene Cutting AKA Oxygen Cutting, Gas Welding or Gas Cutting

This is a process for cutting metal plate in which oxygen and acetylene are combined at the end of a nozzle and ignited, producing a high-temperature flame. The metal is preheated with this mixture of gases, and then a stream of high-purity oxygen is injected into the centre of the flame, which rapidly oxidises the work piece. Because thermal cutting methods are based on a chemical reaction between the oxygen and iron (or titanium), thin or narrow materials are not suited to the process because the heat can cause them to distort.

This sort of cutting can be undertaken either manually or as an automated process. In the manual operation, the familiar worker in overalls, with full-face protection, provides the traditional image that sums up this process. In this scenario, the worker may often be welding, rather than cutting, materials.

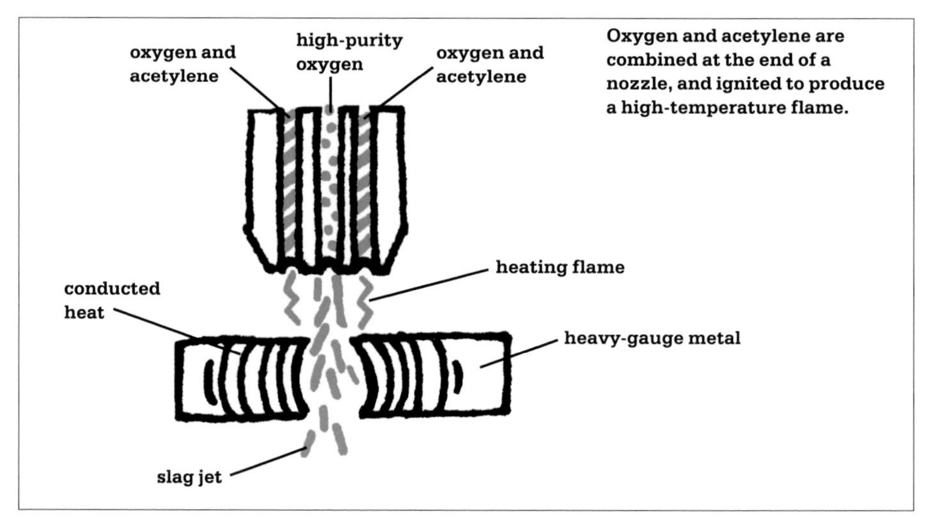

Oxygen and acetylene are combined at the end of a nozzle, and ignited to produce a high-temperature flame.

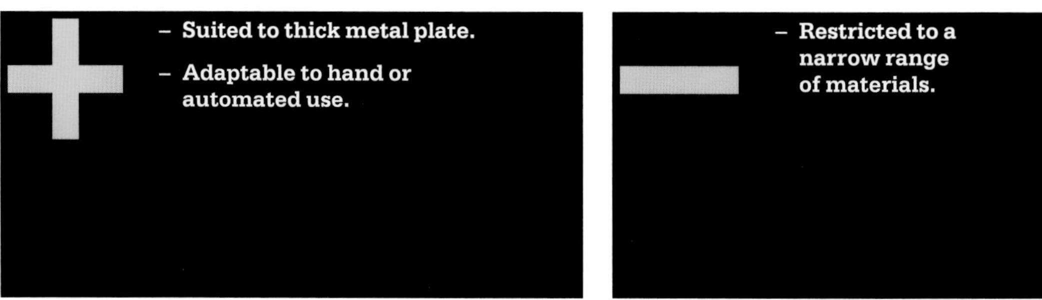

- Suited to thick metal plate.
- Adaptable to hand or automated use.

- Restricted to a narrow range of materials.

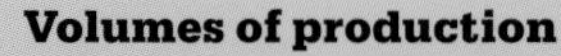

Volumes of production
Compared with the alternatives for cutting thick metals, thermal cutting is an economical process for small-batch production.

Unit price vs capital investment
Unless a cutting template is introduced, the process does not require tooling. In an automated process, the information for the shape can be provided by CAD files. Both of these factors mean that costs are kept down.

Speed
The speed is greatly affected by the type of material used and its thickness. The process may be carried out manually, or it can be highly automated, with multi-torch, computer-operated systems. Speeds can reach up to 3 metres per minute.

Surface
Cutting can be controlled to produce different grades of surface depending on the cost-versus-edge quality – that is, longer cutting times equal better edge finish. The finish of the edge is also determined by the material, but generally plasma-arc cutting (see p.33) will give the best finish.

Types/complexity of shape
The process is best suited to heavy-gauge materials. Metals of below 8 millimetres may distort as a result of the intense heat, as might narrow sections. As in all sheet-cutting operations, nesting one shape within another (as you would when cutting biscuits from dough) to optimise space in between shapes produces an economical use of the material. The cut is generally at 90 degrees to the plate. Other angles can also be achieved, although this is not as easy to set up for oxyacetylene cutting as it is for plasma cutting.

Scale
Using handheld cutting tools there is no maximum size, while in the automated process part sizes are restricted to the size of the machinery.

Tolerances
Depend on the thickness of the material but, as a rule of thumb, they vary between ±1.5 millimetres for 6–35 millimetre-thick materials.

Relevant materials
Limited to ferrous metals and titanium.

Typical products
Heavy construction, including shipbuilding and machine components.

Similar methods
Electron-beam machining (EBM) (p.24), plasma-arc cutting (p.33), laser cutting (p.46) and water-jet cutting (p.42).

Sustainability issues
Oxyacetylene cutting is extremely energy intensive because of the phenomenal temperatures required to maintain the heat of the flame, along with the slow speed at which the torch needs to move, which increases cycle times. In addition, several harmful chemicals are produced from both the fuel and the work piece.

Further information
www.aws.org
www.twi.org.uk
www.iiw-iis.org

Sheet-Metal Forming

Making objects from sheet metal is one of the earliest methods of human production. The Egyptians, for instance, made soft precious metals, such as gold, into sheets, from which they cut sometimes highly intricate forms.

One of the most refined applications for sheet-metal forming can be found in the production of the common whistle. Falling under the generic heading of solid-state forming, the production of whistles is an industrial craft, a multi-stage process that is based on the conversion of a sheet material into a three-dimensional object by cutting, press forming (see metal cutting, p.59) and, finally, plating sheets of brass. However, this overly simplified description masks the fact that this is a precise method of industrial production that requires extremely high levels of tolerance to produce a whistle with the perfect pitch.

The basic geometry of a whistle's body consists of three parts: an underside, mouthpiece and top and side. The pieces of brass are stamped out to form the flat nets and are then pressed into shape using a male and female jig. These components are soldered together, polished and plated in nickel. The final simple step: a cork pea is pushed into the mouthpiece.

In any wind instrument, the sound is the result of air flowing at different rates over a very sharp edge, producing two vibrating columns of air. After 135 years, Acme Whistles, in Birmingham in the UK, has tailored this highly precise process into an art form, producing a reject rate of just 3 per cent for their whistles. Considering the potential for the slightest, sometimes invisible, imperfection to produce the wrong sound, this really is a feat of crafted industrial manufacturing.

Product	Acme Thunderer whistle
Designer	Joseph Hudson
Materials	nickel-coated brass (image shows the brass before plating)
Manufacturer	Acme Whistles
Country	UK
Date	1884

The Acme Thunderer is shown here in its pre-assembled state, and before it is nickel-plated. This shows how many formed components go into making the final product.

Volumes of production
This is a semi-automated method, so it can be used for production runs of greatly varying lengths.

Unit price vs capital investment
This varies greatly, depending on the set-up and the volume of production required. Jewellers can use simple tools requiring very little investment. By contrast, millions of pounds would be needed to set up a production process for the whistle (pictured).

Speed
Varies according to set-up. The Acme Thunderer featured here takes up to three days to produce.

Surface
Generally, this is dependent on the finish of the sheet material, though polishing and painting are often required.

Types/complexity of shape
The nature of this type of set-up allows jigs to be built to accommodate a range of quite complex shapes.

Scale
There is no maximum size for sheet forming.

Tolerances
Can be extremely high. In order to achieve perfect pitch in the whistle, tolerances are ± 0.0084 millimetres.

Relevant materials
Soft metals such as brass, copper and aluminium are particularly easy to form, but any sheet metal can be used.

Typical products
Sheet-metal forming is used to produce a number of products in a variety of industries – they range from brass musical instruments to computer housings and car bodies.

Similar methods
Other processes that give form to flat sheets of metal include metal spinning (p.56), stamping and punching (see metal cutting, p.59), water-jet cutting (p.42), laser cutting (p.46) and CNC folding, a process in which sheet material, usually metal, is folded into different shapes – think of a biscuit tin.

Sustainability issues
As this process is largely automated it consumes a fair amount of energy while the numerous stages of production can increase cycle times. However, excess or metal off-cuts can be recycled back into the process; aluminium is one of the most recycled materials.

Further information
www.acmewhistles.co.uk

- **The beauty of this process is that it allows the creation of a complex form with a highly precise component.**
- **Reasonable tooling costs.**

- **Limited to sheet materials.**
- **The product may have to go through a number of stages.**

Slumping Glass

To slump glass is to allow it to sink into shape. Most people know that if a sheet of glass is left alone long enough, its shape will slowly distort. However, glass does need to be heated to a sufficiently high temperature for it to reach an elastic state that enables it to move at an economical rate or, at the very least, faster than the hundreds of years it would take without heat. When a sheet of stiff glass is placed over a refractory mould (a mould made from a heat-resistant material) in a kiln and heated to 630°C, the glass relaxes enough to allow it to sag into a shape that becomes permanent once cooled.

To form the Fiam table (pictured), a blank sheet of 12-millimetre crystal glass is first cut. The computer-controlled process employs a jet of water, mixed with an abrasive powder, that passes at 1,000 metres per second through a tiny nozzle. This creates a jet strong enough to cut through any material. Once the flat blanks have been cut, the sheet is ready for curving.

The entire sheet and the refractory mould must be brought to the same critical melting temperature – even the smallest temperature variation can result in a broken sheet. At the right temperature, the glass is relaxed enough to sag under its own weight and sink into the mould, with a bit of manual assistance. The apparent simplicity of Fiam's products conceals the complex heating process, which has to be tightly regulated to keep the glass at exactly the right temperature inside the curving chamber. The pieces may be based on simple ideas and shapes, but this simplicity is only achieved (with a high success rate) by the use of sophisticated modern technology.

Product	**Toki side table**
Designer	Setsu and Shinobu Ito
Materials	float glass
Manufacturer	Fiam Italia
Country	Italy
Date	1995

The full radius of the table-top curve, and the gentle curves of the feet, offer a suggestion of the simple forms that are possible with glass slumping.

i

Volumes of production
Slumping is the kind of process that is as well suited to one-offs as it is to batch production.

Unit price vs capital investment
Most commercially available moulds are made of either vitreous clay or stainless steel, but it is also possible to use plaster, cement or even found objects for low-volume production. Depending on the complexity of the shape, the process can involve a high failure rate of finished pieces and, therefore, high unit costs.

Speed
Although this is an industrial process, the speed of forming is quite slow and still requires a considerable use of manual labour.

Surface
Completely smooth glass surfaces can be achieved, as well as textures that can be incorporated into the mould.

Types/complexity of shape
This process works on gravity, so it is possible to achieve any shape that is formed from a flat sheet and has a vertical drape.

Scale
Restricted only by the dimensions of the glass sheet and the kiln that provides the heat.

Tolerances
Due to the difficulty of making glass slump into tight corners, coupled with the expansion of the glass, it can be difficult to achieve tight tolerances.

Relevant materials
Most types of sheet glass (including borosilicate), soda-lime glass and advanced materials such as fused quartz and glass ceramic.

Typical products
Domestic products such as bowls, plates, magazine racks, tables, chairs and tableware. Industrial applications include automotive windscreens, lighting reflectors, furnaces and fireplace windows.

Similar methods
Draping glass over a mould, rather than into a mould, is also a valid process, and is sometimes just called 'draping'.

Sustainability issues
Working with glass is always extremely energy intensive as high temperatures need to be sustained in order to keep it in a workable state. In addition, the shaping is done by hand so faults can often occur. Any rejects from errors or breakage can be recycled back into the process through melting, which requires further heating.

Further information
www.fiamitalia.it
www.rayotek.com
www.sunglass.it

+

- **Allows sheet glass to be formed into a unique three-dimensional shape in as little as a single operation.**

−

- **Slow, and a high degree of skill and experience can be required to trial and error a design.**

Electromagnetic Steel Forming

Electromagnetic pulses in manufacture may seem very complex, but this new use of an existing technology is likely to revolutionize the production of large steel parts for the automotive industry.

At present, large steel sheets are cut using a metal stamping process in which a heavy mould presses the sheet into a die shape. However, this has a number of drawbacks including the enormous size of the machinery and the poor quality of the cutting edge, which is very jagged and requires a secondary finishing process often done by hand. For these reasons electromagnetic-pulse forming is revered as the next big contender in steel forming: it offers improvements in both areas and reduces costs and lead time.

If you try to make two magnets touch, you either feel them attract, or you turn them around and you feel a force that physically pushes them away from each other. This manufacturing process simply amplifies this latter force, using a much greater and carefully directed amount of magnetism by means of a coil, an electrical current and a steel sheet. A capacitor (the container holding the electrical current) discharges the current rapidly through the coil where it is converted into a powerful magnetic field. The pressure of this magnetic field hits a steel sheet placed near the coil and the opposing forces between the sheet and the current in the coil is so strong that it causes the metal to deform. Energy is directed and concentrated so that very precise and refined punches are made into the steel sheet without any physical contact. The impact pressure is said to be the equivalent of three small cars pressing on an area about the size of a fingernail.

The use of electromagnetic pulses is not new – in the past they were used in warfare to disable telecommunications. More recently the technology has been used for small-scale tube forming. To apply it to large sheets of steel, scientists simply altered the machines for tube forming by boosting the coil and the rate at

+

- A very clean and refined punch is achieved, so secondary finishing processes are eliminated.
- The process could eliminate the need for moulds or dies and, in turn, reduce costs significantly.
- The risk of operators being injured is reduced in comparison to stamping.

- The process is still in development so it could be a while before it is commercially viable.
- Energy consumption is likely to be very high because of the high pressures required.

which the charge could be converted. Researchers are working on developing coils that will cut specific shapes and geometries.

Volumes of production
The process is being developed predominantly within the automotive industry for mass-production scales.

Unit price vs capital investment
High investment costs for start-up production. However, existing electromagnetic forming machines can be altered quite simply with a more powerful coil. Overall costs are reduced due to low maintenance of parts and the elimination of secondary finishing processes.

Speed
The process can cut parts up to seven times faster than laser cutting. Speeds are incredibly high (a 30 millimetre-diameter hole can be punched in one-fifth of a second).

Surface
The main benefits of the process derive predominantly from the use of a magnetic field in place of a cutting tool (as with stamping): there is no wear or damage to any parts as no physical contact is made between the materials. As a result, the punched edges are very refined, which eliminates the need for additional finishing and in turn reduces cycle time and costs.

Tolerances
Testing has shown that the process can punch holes in stainless steel and other hardened metals, and could also be used to form shapes in metal without the need for a die or mould. This opens up entirely new opportunities within the manufacture of heavy metals, affirming the process as a valuable tool in the future of automotive and vehicle production.

Relevant materials
A range of magnetic metals (that is, the process is not suitable for use with aluminium etc.), and significantly tough materials such as stainless steel and other hardened metals.

Typical products
The process is best suited to large panels, such as car doors, frames and bonnets. It is also used for home appliances, including washing machines, dishwashers and fridges, etc.

Similar methods
Superforming aluminium (p.70), press forming (p.59), Industrial Origami® (p.61).

Sustainability issues
Energy consumption is likely to be very high due to the intensive pressures that are required to deform the steel sheets. However, electromagnetic forming requires no tooling or contact with the steel, which reduces material consumption through maintenance and replacement of parts during the lifespan of the machine.

Further information
http://www.fraunhofer.de/en

Metal Spinning
including sheer and flow forming

Spinning is a widely used technique for bending sheet metal. As the name suggests, the process involves a flat metal disc, known as the blank, being spun, pushed and consequently wrapped around a rotating mandrel to produce curved, thin-walled shapes. A flat metal sheet (the blank) is first clamped against the mandrel and then both are rotated at high speed, in the same direction. The spinning metal is then pushed with a tool – which in hand spinning is sometimes called the 'spoon' – against a wooden mandrel until it fits to the mandrel's shape. The resulting part is thus a copy of the

Product	**Spun**
Designer	Thomas Heatherwick
Materials	brushed/polished steel or copper
Manufacturer	Haunch of Venison
Country	UK
Date	2010

These large metal spinnings perfectly illustrate the typical forms that are the result of the process, even down to the spinning lines which are clearly visible. Though these works are substantial, even larger forms are possible.

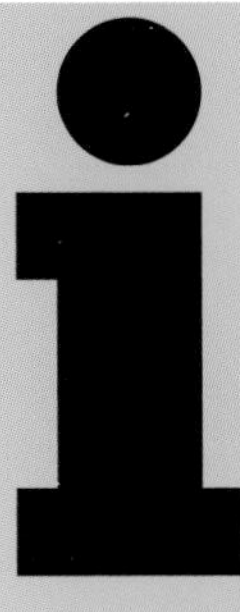

Volumes of production
From single prototypes to batch production and runs of several thousand.

Unit price vs capital investment
The tools for pushing and the mandrels are made from wood or metal, depending on the size of the component and the quantity required. For a small number of units it makes sense to use affordable wooden mandrels, but for large production runs metal is a better choice because the former will be subject to greater wear.

Speed
Production cycle times are higher than for press forming (see metal cutting, p.59), but set-up times are substantially shorter, making metal spinning suitable for prototyping, one-offs and short- to medium-batch production.

Surface
A spun surface may need to be polished in order to eliminate the circular witness lines on the external surface of the part.

Types/complexity of shape
This is really only a technique for making symmetrical shapes that start with a sheet of metal. Discs, cones, hemispheres, cylinders and rings are the typical shapes made using this process. Undercuts and re-entrant angles are achieved by a split mandrel, which comes apart like the segments of an orange in order for the part to be released. Closed shapes such as hollow spheres are made by joining two halves together.

Scale
Spun-metal products can be produced at less than 10 millimetres in diameter, and, at the other end of the scale, Acme Metal Spinning in the US has produced a shape that measures almost 3.5 metres across.

Tolerances
Because the metal is stretched around the mandrel, the thickness of a part changes during the spinning process. The flatter the shape, the less the metal will need to stretch.

Relevant materials
Spinning can be applied to a variety of metals, ranging from soft, ductile coppers and aluminium (which are the most common) to hard stainless steels.

Typical products
The kitchen wok is a good example of an item made by spinning – it is even possible to see the evidence of its production in the concentric lines on the outside surface. Other products include bases and lampshades for lighting, cocktail shakers, urns and a whole mass of industrial components.

Similar methods
Spinning is often combined with other techniques to produce more complex products. For example, pressure-formed parts are often spun to create necks, flanges and flares. Although far less common than spinning, incremental sheet-metal forming (p.257) is a new process that allows a range of complex forms to be created from sheet metal using a single tool.

Sustainability issues
Metal spinning is one of the most energy-intensive processes available as the part needs to be constantly rotated at a high speed for a long length of time. However, with the formed metal's high strength, parts have excellent durability which ensures a prolonged life cycle. Additionally, at the end of the product's lifespan the metal can be melted down and reused.

Further information
www.centurymetalspinning.com
www.acmemetalspinning.com
www.metalforming.com
www.metal-spinners.co.uk

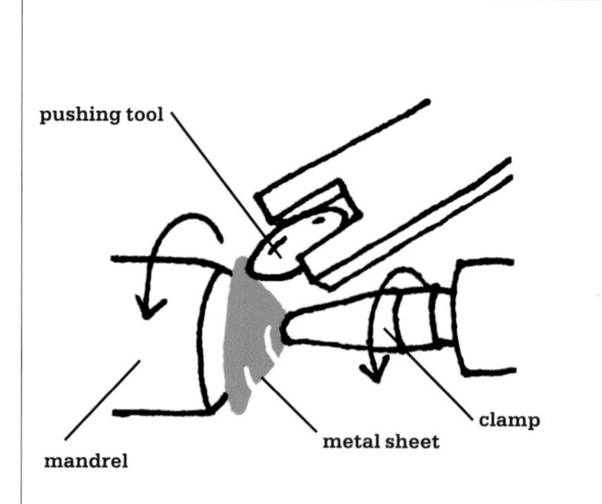

A flat metal sheet is clamped against a mandrel and both are rotated at high speed. The metal is pushed with a tool until it conforms exactly to the mandrel's shape.

external shape of this mandrel. Several operations can be performed in one set-up, and work pieces may have re-entrant profiles (undercuts). The design profile in relation to the centre line is therefore virtually unrestricted, though it will be symmetrical.

Sheer and flow forming are advanced forms of spinning that can be used to deliberately alter the wall thickness of metal parts by anything up to 75 per cent. It is ideal for concave, conical and convex hollow parts.

1 Preparation of the wooden mandrel.

2 The metal is pushed against the mandrel as both metal and mandrel are spinning.

3 The metal component taking shape over the mandrel.

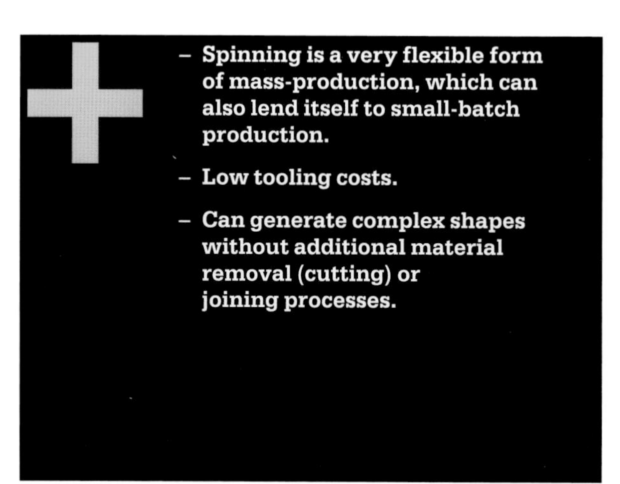

- Spinning is a very flexible form of mass-production, which can also lend itself to small-batch production.
- Low tooling costs.
- Can generate complex shapes without additional material removal (cutting) or joining processes.

- Some materials will harden in the spinning process.
- Spinning often requires post finishing.
- The process offers limited control over wall thickness because of the way the metal is slightly stretched over the mould.

Metal Cutting

including press forming, shearing, blanking, punching, bending, perforating, nibbling and stamping

In the metal industry the term 'cutting' is hardly ever used, because technically it is such a broad term it has almost no meaning. Cutting processes can be divided into two main categories: chip-forming and non-chip-forming. Press forming, shearing, blanking, punching, bending, perforating, nibbling and stamping are all terms that in one way or another describe non-chip-forming of metal sheet. Methods such as milling (see machining, p.18) and turning on a lathe (see p.26), on the other hand, are chip-forming techniques.

Punching and blanking are very similar in the sense that they both involve the removal of part of a sheet to form a hole. The processes differ in that punching is used to make sheets with shapes cut out of them, while blanking is a process for making separate shapes, similar to using a biscuit cutter to make many biscuits from rolled-out dough. The metal disc, which is the starting point for the beverage can tops (pictured), would have been made using blanking.

Nibbling is used to cut a sheet in successive bites from a small punch that pulses up and down in a process similar to that of a sewing-machine. Shearing involves a punch and a die with a tight control over the gap between the two (unlike punching, which doesn't have a die). The terms 'perforating' and 'bending' should be fairly self-explanatory.

Product	beverage can ring-pull
Materials	aluminium
Manufacturer	Rexam
Country	UK
Date	1989

This is an everyday product that has to be super cost-effective, yet must work all the time and must absolutely never cut your lip when you drink from it. Press forming and shearing are just two of the methods used to make this ubiquitous product.

Metal stamping is a cold forming process that is used to produce shallow components from metal sheet. Although it is a fairly straightforward method of cutting and forming sheet, it includes several variations, all of which combine a punching process together with a forming process, performed either in a sequence or in one action. A single die is needed for each operation, but the component can be removed and placed in another die for additional forming. Progressive dies (like a series of dies) are used in more complex procedures to form multiple actions.

Volumes of production
The process can be used for manual production or for an automated CNC high-volume production.

Unit price vs capital investment
Tooling costs can be reduced, or eliminated, by the use of existing punches or cutters, allowing for high-volume production to be achieved with low capital costs.

Speed
Varies greatly, but typically 1,500 drinks-can ring-pulls can be produced per minute.

Surface
In terms of finishing, these cutting techniques will generally need deburring.

Types/complexity of shape
Mostly used in the production of small components, and thickness is restricted to available standard sheet.

Scale
Restricted by the standard sheet dimensions.

Tolerances
High tolerances are achievable.

Relevant materials
Restricted to sheet metal.

Typical products
Cooling fan blades for electronics, washers, keyholes and watch components.

Similar methods
Laser cutting (p.46) and water-jet cutting (p.42) are two non-chip-forming methods that can be set up to produce designs from CNC programs, without tooling costs.

Sustainability issues
Each of the various cutting processes is based on the removal of material, which results in a significant amount of waste material. However, metals can be melted down to form new sheets that can reused in the process to reduce material consumption and the use of virgin resources. Aluminium is one of the most widely recycled materials.

Further information
www.pma.org
www.nims-skills.org
www.khake.com/page88.html

+

- **Very versatile in terms of producing different shapes.**
- **Can be used for any solid metal.**
- **High degree of accuracy.**

- **Parts may be limited to stock sizes of material.**
- **Material utilisation can be low due to wastage.**

Industrial Origami®

There is something fascinating and inspiring about watching a simple, flat sheet of paper being transformed into a complex form through origami. In much the same way, this patented process takes the principles of origami but applies them on a much more industrial scale, using metal in place of paper to create usable products.

This folding innovation has many benefits over traditional metal-forming methods such as stamping and press breaking, as it reduces the number of operations required to shape the metal and the whole process can therefore be completed in much less time and at a much lower cost. The component is created from a net, much like a flattened cardboard box. A stamping technique or a laser is used to cut the outline of the shape from a metal sheet and to produce a series of lines and smile-shaped curved cuts along the edges to be folded. A set of straps pulls at the smile shapes from either side of the sheet,

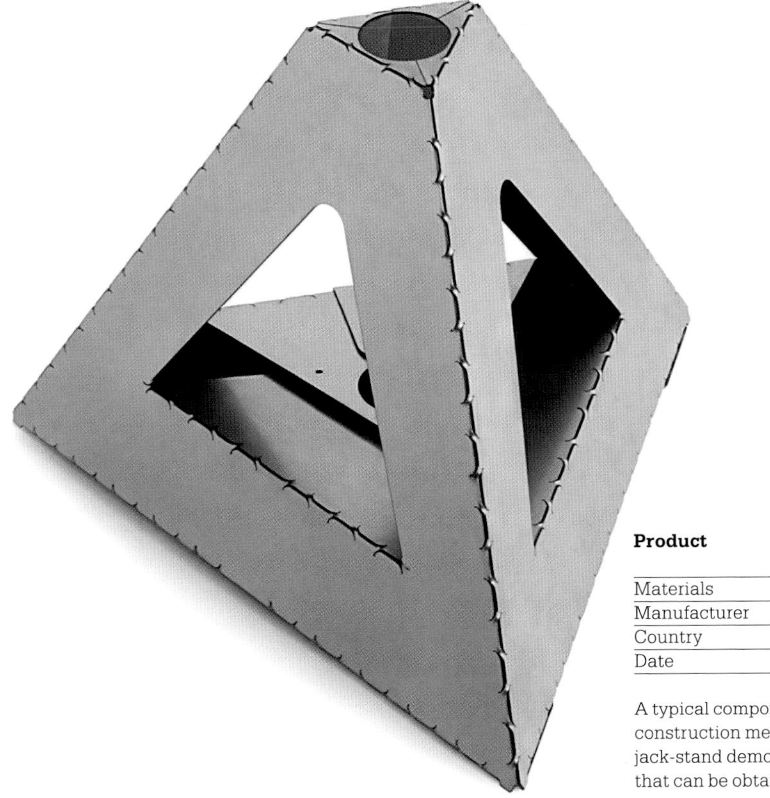

Product	Jack-stand made with Industrial Origami®
Materials	12-gauge cold-rolled steel
Manufacturer	Industrial Origami
Country	USA
Date	2004

A typical component showing the cutlines and construction method of Industrial Origami®. This jack-stand demonstrates the structural strength that can be obtained using the method.

to create contact between two of the sides. This leverage causes the sheet to bend along the fold lines with only a relatively small amount of force. It is the small smile-shaped cuts that control and determine the folds as they direct the stresses during folding to make everything align perfectly.

The process allows for the integration of several parts into a single piece, and removes the need for welding and joining as it uses a number of folding clips to secure the folds, which significantly reduces material consumption. The process enables the rapid creation and fold-up of prototypes, which allows designers to experiment with and test prototype configurations quickly, and make any necessary changes.

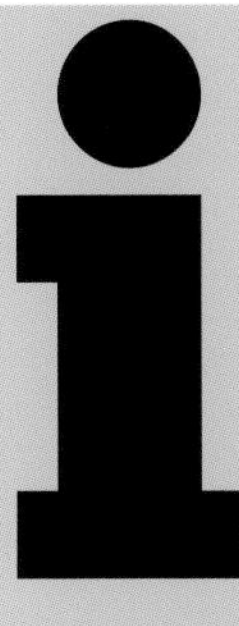

Volumes of production
From one piece to millions of pieces.

Unit price vs capital investment
Investment costs of designing prototype productions and for manufacture are high. However, savings on materials, storage and transportation, and the elimination of secondary fixing processes make the process cheaper overall.

Speed
Origami features are dependent on the available punch, laser or stamping speeds. Folding is complete in seconds.

Surface
Not applicable.

Types/complexity of shape
Suitable for a range of thicknesses from 0.25 millimetres to the limit of any cutting equipment.

Tolerances
Alternative methods of forming in metal can produce stack-up errors related to maintaining dimensional tolerance. Industrial Origami maintains the level of accuracy of the machine that originally applied the lancing.

Relevant materials
The process has been predominantly used with sheet metals but it could be applied to any number of materials, including plastics and composites.

Typical products
Currently the main applications are for engendering components such as automotive chassis systems, solar mounting systems, packaging, cooker tops and built-in ovens.

Similar methods
Superforming aluminium (p.70), press forming (p.59).

Sustainability issues
The sheets are stamped in flat pieces so can be transported flat-pack and folded on arrival. This reduces the space needed during transportation, which could potentially allow for greater effectiveness of energy use. Additionally, there is a significant reduction in materials consumption as many joints and fixtures are incorporated into the fold design.
As with any sheet-cutting process there is a high degree of waste materials; however these can potentially be recycled.

Further information
www.industrialorigami.com

+

- **Reduced joining, fixing and processing.**
- **Integrating multiple parts into a single sheet means material consumption is significantly reduced.**
- **Fast construction and assembly relative to alternative methods of construction.**
- **Allows for effective prototype testing.**
- **Lower labour costs.**

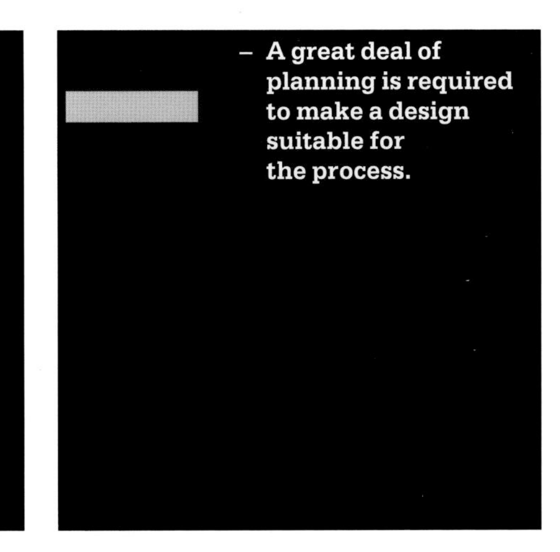

–

- **A great deal of planning is required to make a design suitable for the process.**

Thermoforming

including vacuum, pressure, drape and plug-assisted forming

Thermoforming is one of the most common methods of producing plastic components and any art student who has been through the traditional foundation course will have used a vacuum-forming machine. Vacuum forming is one of the few plastic-forming methods that is as accessible to school and college students as it is to large-scale industrial production. It is also one the easiest methods of production to comprehend, and if you have ever seen the process in operation you will understand why.

The basic materials needed for this process are a thermoplastic sheet and a former. Because the pressures employed are low, the former can be made of wood, aluminium or other fairly inexpensive materials. The former is the exact shape of the part required and is placed at the centre of a table, which can be raised and dropped. The rigid plastic sheet is heated under a series of convection bars similar to a domestic oven, until the plastic is soft, pliable and saggy. At this point, the former is raised on its bed, pushing into the soft sheet, and a vacuum is applied. This sucks the air out from below and pulls the plastic onto the former. Once the plastic has 'hugged' the former and cooled slightly, it can be removed for post finishing.

Product	**chocolate-box tray**
Materials	Plantic biodegradable polymer
Date	2005

There are few better examples of thermoforming than a chocolate-box tray. The individual shapes of the chocolates are evidence of the shape of the mould that is used to form the trays.

Other types of thermoforming include pressure forming, which works in the opposite way to vacuum forming by forcing the material into the mould. Drape forming, as the name suggests, consists of draping a sheet of heated plastic over a male mould, where it is mechanically stretched allowing the sheet to remain close to its original thickness. Plug-assisted forming uses plugs to pre-stretch the plastic before the vacuum is introduced. Again, this allows for greater control over the material thickness.

Volumes of production
Suitable for model-makers' prototype work and one-offs, but also for large-scale production.

Unit price vs capital investment
Formers can be made from a range of materials, depending on the number of components that is required. The ease of machining and its wear-resistance makes aluminium suitable for large production runs. Epoxy resins are used as a cheap alternative to aluminium, but anything can be used, including MDF, plaster, wood and even plasticine, which is actually excellent for vacuum-forming shapes with undercuts, because you can pick it out afterwards.

Speed
Bathtubs can be made at a rate of one every five minutes. Beyond this, speed is difficult to estimate, because multi-mould formers can rapidly speed up the process, and, besides this, the thickness of the material affects the time it takes to heat up sufficiently.

Surface
Vacuum forming picks up surface details very well, so the surface finish of the mould is reflected in the surface finish on the part.

Types/complexity of shape
You will need draft angles, because undercuts are impossible to achieve with standard tooling.

Scale
A 2 by 2-metre aperture is standard, but it can be even bigger.

Tolerances
Varies, depending on size of the formed piece. As a guide, a forming of less than 150 millimetres will hold a tolerance of 0.38 millimetres.

Relevant materials
Most thermoplastics that are supplied as sheets. Typical examples include polystyrene, ABS (acrylonitrile butadiene styrene), acrylics and polycarbonates.

Typical products
Canoes, bathtubs, packaging, furniture, interior car trim and shower trays.

Similar methods
Superforming aluminium (p.70) and inflating metal (p.76).

Sustainability issues
The low pressure and moderate temperatures employed during processing, along with fast cycle times, ensure that energy consumption is low. However, additional processing is required to trim the excess material and this produces a significant amount of waste plastic. Further heat is needed for these offcuts to be melted down and recycled. The nature of the shapes that are thermoformed means components can be nested during transportation, saving on bulk.

Further information
www.formech.com
www.thermoformingdivision.com
www.bpf.co.uk
www.rpc-group.com

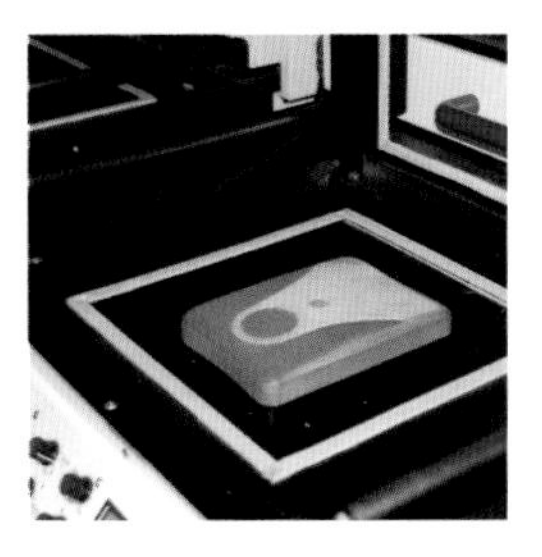

1 The former (in this case a simple wooden shape for a college project) is placed on the bed.

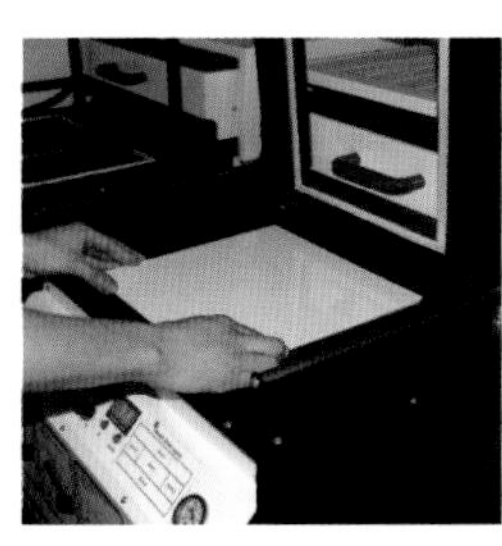

2 Once the bed has been lowered, the plastic sheet is placed on top, ready to be clamped by the metal frame.

3 The heater is lowered onto the plastic sheet.

4 A vacuum is applied to form the shape.

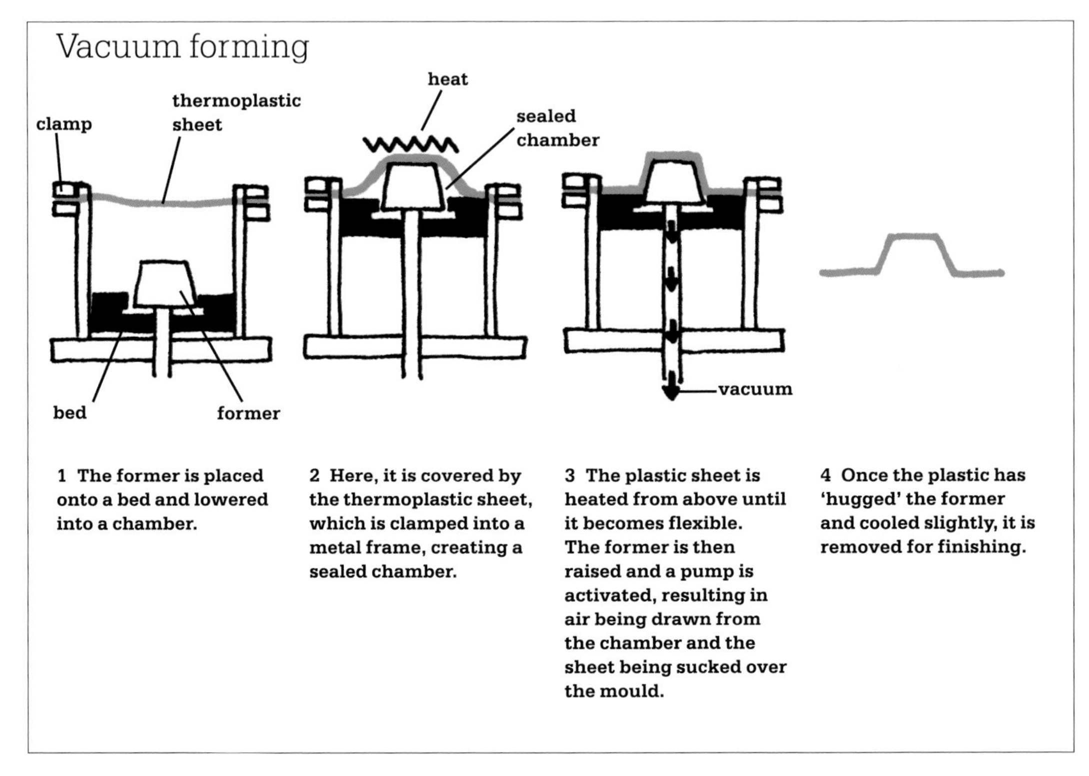

1 The former is placed onto a bed and lowered into a chamber.

2 Here, it is covered by the thermoplastic sheet, which is clamped into a metal frame, creating a sealed chamber.

3 The plastic sheet is heated from above until it becomes flexible. The former is then raised and a pump is activated, resulting in air being drawn from the chamber and the sheet being sucked over the mould.

4 Once the plastic has 'hugged' the former and cooled slightly, it is removed for finishing.

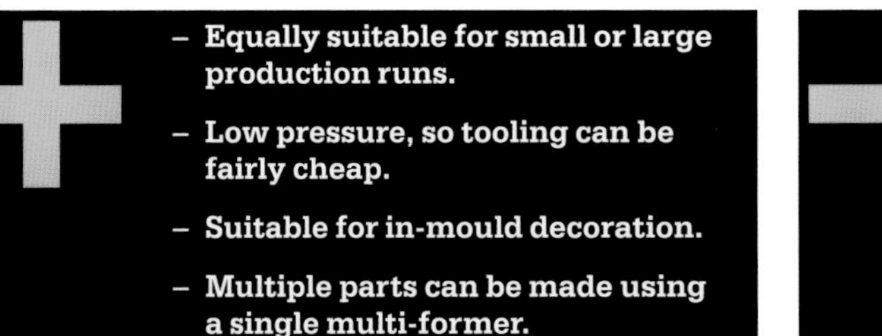

+

- Equally suitable for small or large production runs.
- Low pressure, so tooling can be fairly cheap.
- Suitable for in-mould decoration.
- Multiple parts can be made using a single multi-former.

–

- Requires a secondary process to trim the sheet.
- No vertical sides on the finished part; draft angles are a must.
- Can have undercuts, but these need special tooling.

Explosive Forming AKA High-Energy-Rate Forming

Can you imagine the fun in discovering this process? In a way it reminds me of the TV-show character Mr Bean, who decided to paint his living room by planting an explosive in a tin of paint. However unlikely, explosive forming is actually an established method of forming metal sheet or tube. It is also another great example of a process that demonstrates the lateral thinking used by engineers to pursue new methods of making things.

The first use of explosive forming was documented in 1888, when it was employed in the forming of plate engravings. The First and Second World Wars provided an intense period of development, as a result

Product	Desert Storm architectural panels
Materials	coil-coated aluminium
Manufacturer	3D– Metal Forming BV
Country	The Netherlands
Date	1998

These architectural panels show the scale of panels and the complex patterns that can be achieved with explosive forming.

of which explosive forming became a major process for manufacturing missile nose cones in the 1950s. Today, explosive forming exists in two forms – 'standoff', in which the explosive is positioned at a distance from the metal, either in the open air or submersed in water or oil, and 'contact forming', in which the explosive is in direct contact with the metal.

In simple terms, the sheet or tube is placed in a vacuum-sealed die cavity, which is in turn placed under water (unless it is the open-air method). A charge is placed over the sheet and detonated, sending shockwaves through the water and rapidly forcing the material into the die cavity.

Although this image does not show the close-up workings of the process, it does give some indication of the scale and the sealed, pressurised environment in which the explosive forming takes place.

+

- It is possible to achieve precise tolerances.
- Cost-effective tooling compared to alternatives.
- Can reduce the number of operations in the manufacturing process, including welding, due to its ability to form complex parts.

–

- Limited number of manufacturers.
- Must adhere to strict safety regulations.

i

Volumes of production
Explosive forming can be used for one-off art projects such as sculptures and installations, but it is equally suitable for mass-production of industrial components. In former East Germany it was used to make hundreds of thousands of cardan axles for heavy trucks.

Unit price vs capital investment
If conventional pressing or spinning can be used, they would usually be cheaper, but relatively low tooling costs and the ability to manufacture complex shapes can make explosive forming the best option available.

Speed
Varies enormously depending on the size and complexity of the shape. Sometimes it is possible to manufacture twenty small parts in one explosion, while larger, more intricate shapes can require up to six explosions over three days. Even a single explosion is quite time-consuming, however, due to the lengthy set-up time (amounting to over an hour per explosion).

Surface
Surface quality is generally extremely good. It is possible to form grade 2G (chemically polished) stainless steel without damaging even the protective foil, producing parts with a perfect mirror finish.

Types/complexity of shape
Ideal for forming complex shapes with seamless cavities.

Scale
Specific manufacturers can form sheets of nickel up to an incredible thickness of 13 millimetres, with lengths of up to 10 metres. Larger sheets are only achievable by welding sheets together.

Tolerances
Able to maintain precise tolerances.

Relevant materials
The process is not restricted to soft metals such as aluminium, but embraces all metals, including titanium, iron and nickel alloys.

Typical products
Large architectural components and panels, and parts for the aerospace and automotive industries.

Similar methods
Superforming aluminium (p.70) and inflating metal (p.76).

Sustainability issues
Relatively slow cycle times coupled with intensive energy consumption hinder the use of this process for sustainable manufacture. In fact, some larger forms can require several explosions to deform fully, which further increases energy use. Harmful substances are used to create the explosive chemical reaction and need to be cleaned before disposal.

Further information
www.3dmetalforming.com

Superforming Aluminium

including cavity, bubble, back-pressure and diaphragm forming

The process of heating a sheet of plastic, draping it over a mould and sucking the air out has been in use for some time (see thermoforming, p.64). However, as the speed of the development of new materials increases, more technologies overlap when it comes to both materials and processes. Superforming involves such an overlap, since it brings traditional vacuum forming with plastic to aluminium alloys. The process is achieved through four main methods: cavity forming, bubble forming, back-pressure forming and diaphragm forming, each suited to specific applications. The common element in all these methods is the heating of an aluminium sheet to 450–500°C in a pressurised forming oven, and then forcing it over, or into, a single surface tool to create a complex three-dimensional shape.

Product	**MN01 bike**
Designer	Marc Newson
Frame builder	Toby Louis-Jensen
Materials	aluminium
Manufacturer	Superform Aluminium
Country	UK
Date	1999

This bike is a good example of the transfer of industrial manufacturing processes into consumer products by experimental projects. The text embossed onto the frame also illustrates the detail that is achievable.

In the cavity method, air pressure forces the sheet up into the tool in a process that can be described as 'reverse vacuum forming'. According to the manufacturers, this process is ideal for forming large, complex parts such as automotive body panels.

In bubble forming, the air pressure forces the material into a bubble. A mould is then pushed up into the bubble and air pressure is applied from the top, forcing the material to conform to the shape of the mould. Bubble forming is suitable for deep and relatively complex mouldings that are difficult to achieve with the other superforming processes.

Back-pressure forming uses

Cavity forming

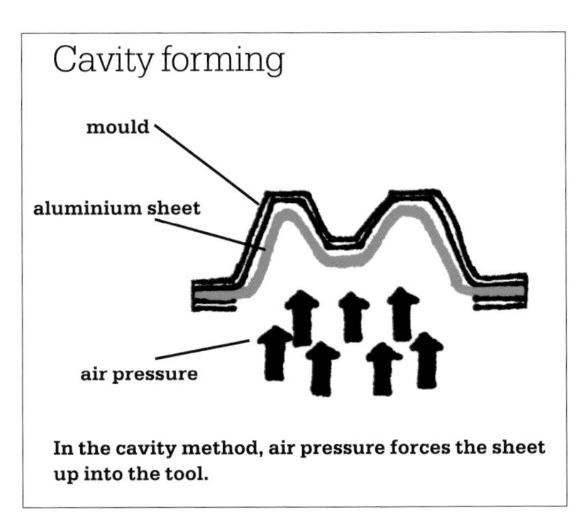

In the cavity method, air pressure forces the sheet up into the tool.

Bubble forming

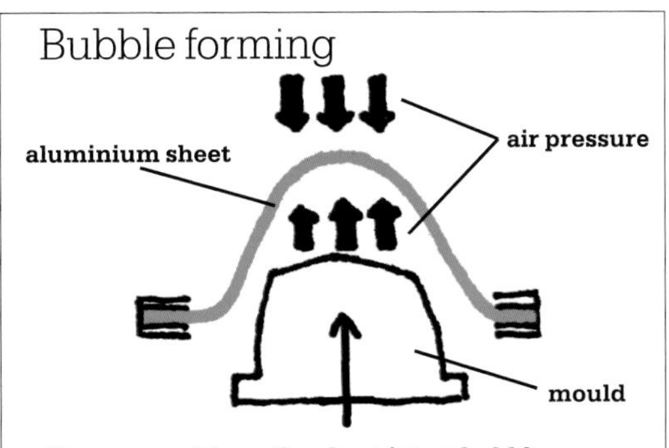

Air pressure blows the sheet into a bubble. A mould is then pushed up into the bubble and air pressure is applied from the top, forcing the material to conform to the shape of the mould.

Back-pressure forming

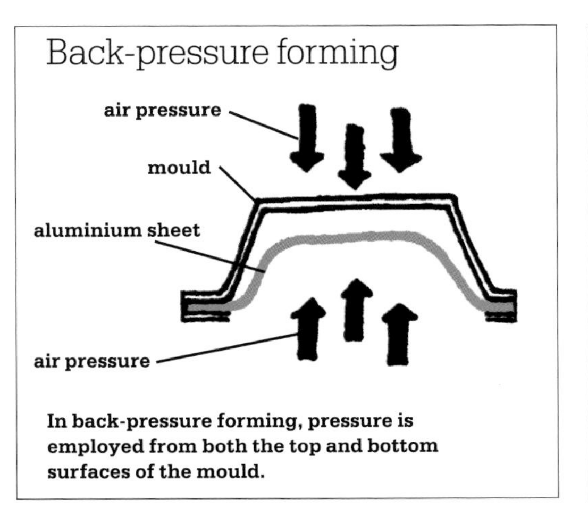

In back-pressure forming, pressure is employed from both the top and bottom surfaces of the mould.

Diaphragm forming

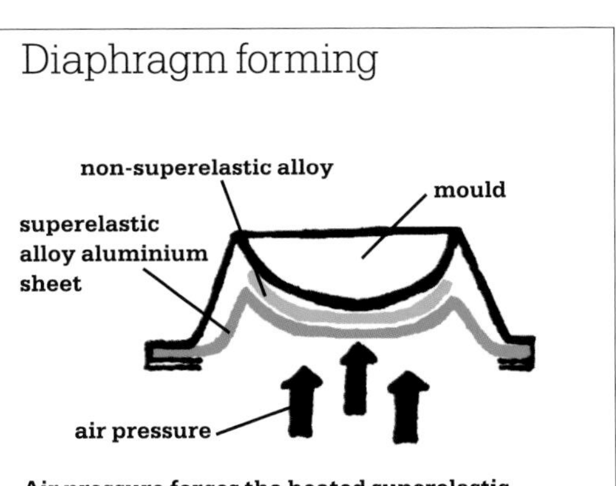

Air pressure forces the heated superelastic aluminium onto a heated non-superelastic alloy which is then formed over the mould.

+

- Complex forms can be created within a single component.
- A range of sheet thicknesses can be used.
- Can create subtle details and forms, without spring-back issues.

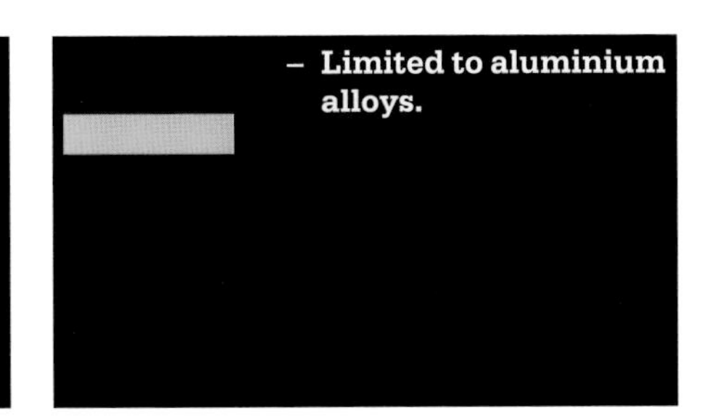

- Limited to aluminium alloys.

pressure from both the top and bottom surfaces of the mould to maintain the integrity of the sheet and allow for the forming of difficult alloys.

Diaphragm forming is a process that allows for 'non-superelastic' alloys to be formed. The non-superelastic material is 'hugged' over the mould using a combination of a sheet of heated 'superelastic' aluminium and air pressure.

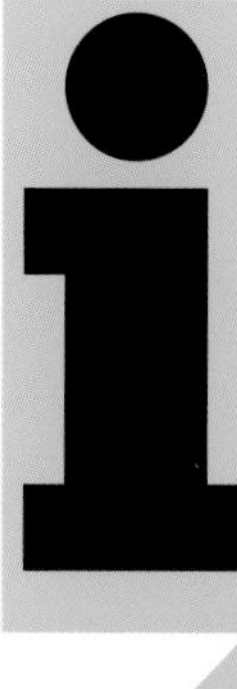

Volumes of production
At present, production runs of about 1,000 parts are considered large, but mass-production is a possibility, with some car manufacturers starting to use the process on a larger scale.

Unit price vs capital investment
High capital investment, mainly in tooling and material.

Speed
Depends on the material – some alloys can be formed in three to four minutes, while the structural alloys used in aircraft, for example, may need up to an hour to form.

Surface
Excellent surface quality.

Types/complexity of shape
This depends on the specific method you use. Bubble forming allows the greatest degree of complexity in shape, but with all methods the basic principle is about creating three-dimensional shapes from a flat sheet. Draft angles need to be considered in order for parts to be removed from the mould. Undercuts are not recommended.

Scale
Each method is suited to different scales and thicknesses of material, for example, using back-pressure forming, parts can be made up to approximately 4.5 metres square. Cavity forming can only process smaller sheet sizes, although these can be up to 10 millimetres thick.

Tolerances
Typically ±1 millimetre for larger parts.

Relevant materials
This process is specifically designed for use with what are known as 'superelastic' types of aluminium. However, the diaphragm-forming method enables the processing of non-superelastic materials.

Typical products
A large market for this process is in the aerospace and automotive industries. Designers such as Ron Arad and Marc Newson have applied it to diverse furniture and bicycles. On the London Underground, architect Norman Foster used superforming to produce tunnel-cladding panels for Southwark station.

Similar methods
For plastic, vacuum forming (p.64), for glass, slumping (p.52), and for metal, look at the inflated stainless steel by Stephen Newby (p.76).

Sustainability issues
The process requires several stages of production each of which uses significant amounts of energy through high heat and pressures. A significant amount of excess material is produced after trimming but can be recycled back into the process or used elsewhere. Additionally, when the formed product reaches the end of its lifespan it can be recycled into new products to reduce the use of raw materials. The nature of the shapes that are thermoformed means components can be nested during transportation, saving on bulk.

Further information
www.superform-aluminium.com

Free Internal Pressure-Formed Steel

Product	**Plopp Stool**
Designer	Oskar Zieta
Materials	stainless steel
Manufacturer	Oskar Zieta
Country	Switzerland
Date	First exhibited 2009

The visual language of these stools refers to the soft forms typical of inflated latex, masking the fact that they are made from stainless steel.

This whimsical and playful process is a buoyant example of a traditional technique used in an unconventional way to produce creative and surprising results. It works like inflatable pool toys or armbands, but uses steel in place of plastic to create objects that are incredibly solid and strong. Oskar Zieta, the designer behind the process, calls it 'FIDU', which stands for free inner-pressure deformation (the letters correlate in his native German).

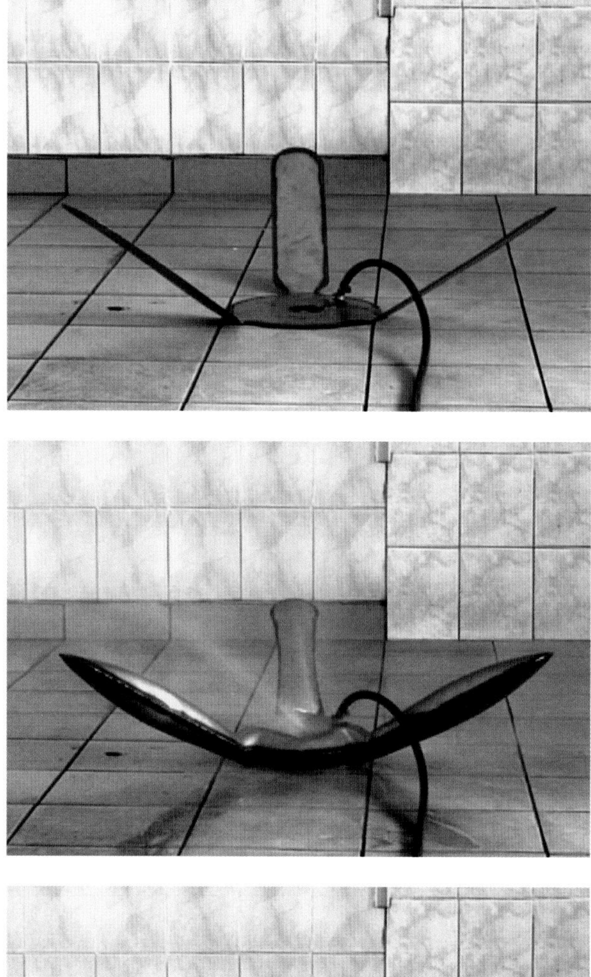

Two sheets of steel are laser cut to produce two pieces that are identical in shape. These pieces are then welded together along the edges by a robot to provide a water- and airtight seal. As the air enters, the sheets begin to deform and expand into a 3-D shape. The result is a light construction, which can be easily mass-customized at low production costs with well-established techniques.

This process, with Oskar's many other sheet-metal-forming techniques, opens up new possibilities for lightweight structures and products that are incredibly strong and stable.

Volumes of production

This is limited to batch production from a single company. Oskar Zieta's creative process involves a hands-on approach where a great amount of experimentation and prototyping is carried out with different shapes to test the behaviour and possibilities of the steel. Therefore this is a process suited to one-off or large-scale batch production.

Unit price vs capital investment

Laser technology is still expensive but the difference between price of the material per kg and laser generator has changed dramatically in the past five years. Cheaper lasers will mean that this technology will be used more frequently in future. The numerous prototyping and product test steps require a lot of set-up time to develop a design, so both lead times and costs are increased; however, in many cases tooling is kept to a minimum.

Speed

It takes 21 minutes to make one of the stools.

Surface

All of the common surface treatments – polishing, powder-coating, lacquering, enamel-coating, gumming – can be used to treat the surface.

Scale

Steel-sheet coils are up to 4km long, the largest non-standard sheets are 3 x 30m. One of Oskar's projects for the Victoria and Albert Museum was 30m long.

Tolerances

The tolerance is dependent on the complexity and the geometry of the shape.

Relevant materials

Sheet steel and plastics.

Typical products

The process had predominantly been used to produce furniture, including stools, chairs and benches. However, this innovative technique has now found its way into the development of wind turbine rotor fins, structural forms such as bridges, exhibition pieces and even bike frames.

Similar methods

Superforming aluminium (p.70), impact extrusion (p.146), press forming (p.59).

Sustainability

Material consumption is low as the steel structures are hollow yet stable and strong. Apart from the sealing of the metal edges no heat is required.

Further information

www.zieta.pl
www.nadente.com
www.blech.arch.ethz.ch

+

- **Produces comparatively lightweight structures from a rigid thin material**
- **Products are highly customisable.**

–

- **There are several stages of production (laser cutting, welding, forming), which increases lead times.**

Inflating Metal

From its extraction when creating vacuums to its sudden introduction into a pre-form to create plastic bottles, air is often a key material in production methods. In terms of blow forming, it is thousands of years old, and its earliest use was in the forming of glass. British designer Stephen Newby, however, has recently introduced a way of inflating stainless steel sheet to create new possibilities in visual language for this hard metal.

The soft appearance of the inflated shapes contrasts with the tough, hard quality of the steel. The process literally involves inflating two sheets of metal that have been sandwiched together and sealed at the edges, without using moulds. Each inflated piece, therefore, responds in a different way, producing a unique piece. In terms of size, the pieces are only limited by the original sheet size. A variety of textured and coloured stainless steels can be used – these are not damaged in the process because the metal is formed from the inside.

Product	**inflated stainless steel pillows**
Designer	Stephen Newby
Materials	stainless steel
Manufacturer	Full Blown Metals
Country	UK
Date	2002

The indentations in these pillow shapes are the natural result of the metal creasing when it is inflated from the two sheets of steel that are sandwiched together.

Volumes of production
Best suited to batch production.

Unit price vs capital investment
No tooling, but some designs require prototyping.

Speed
Blow forming (of which inflating metal is just one example) is instantaneous. The process is semi-automated and runs at different times according to size. For example, an inflated 10-centimetre metal square can be produced at the rate of 30 squares per hour.

Surface
Full range of high-quality factory-applied finishes, including mirror finishes, colours, etchings, textures and embossed finishes.

Types/complexity of shape
Any shape that can be produced from flat two-dimensional templates, including organic forms, figurative lettering, soft, cushion-like creased forms and smooth, uncreased forms.

Scale
From 5 centimetres up to the maximum single sheet size, typically 3 by 2 metres.

Tolerances
5 millimetres per 1,000 millimetres in overall dimensions.

Relevant materials
Most metals including stainless steel, mild steel, aluminium, brass and copper.

Typical products
Architectural cladding and screens, large-scale public art, outdoor design, including water features, and contemporary interior products.

Similar methods
Glass blowing by hand (p.116) and superforming aluminium (p.70).

Sustainability Issues
Although very hard and rigid, only very thin sheets of steel are used to make these steel structures, which reduces material consumption without compromising strength. However, the extreme temperatures required for the metal welding, along with the high pressures required to pump in the air, consume a fair amount of energy.

Further information
www.fullblownmetals.com

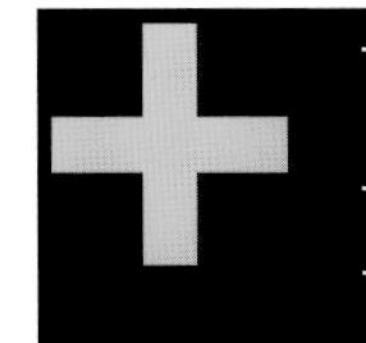

- **Ability to form unique shapes in metal.**
- **High strength-to-weight ratio.**
- **The process can be used to form high-tensile strength materials.**
- **Factory-applied finishes are preserved in the forming process.**
- **Specific dimensions are easily achieved without the need for moulds or jigs.**

—

- **Offered only by a single manufacturer.**

Pulp Paper

Paper can be a wonderfully low-tech material and it's so common, everyday and often banally dull, that we often forget its contribution to life simply because of its sheer number of manifestations.

Evolved from Swedish company Sodra who produce timber-based materials, Pulp Labs are creating a whole range of pulp-based materials and experimenting with manufacturing processes to create some exciting new outcomes. This is a high-tech approach to a low-tech material for creating semi-durable papier mâché products. The process is based on various methods; in one the press forms layers of paper together that have been impregnated with Polylactic acid (PLA), a biodegradable material made from maize starch or sugar cane. Just think of a trouser press, with heat and pressure combining to melt the PLA and bind all the layers of paper together to produce remarkably strong parts. Depending on the geometry of the part, compressed pulp just 1.8mm thick is able to support the weight of a person. Due to the fibres being encapsulated by the PLA the material is also unaffected by humidity and heat, and its durability is almost comparable to that of steel, wood or plastic.

Product	**Parupu**
Designer	Claesson Koivisto Rune
Materials	paper pulp
Manufacturer	Sodra Pulp Labs
Country	Sweden
Date	2009

The Parupu chair combines strong geometry with advanced use of an archaic material. The mould of the chair is dipped into a paper-based slurry, heated in an oven and then pressed to activate the DuraPulp.

Volumes of production
At the time of writing the process is more a laboratory project than one for mass-production. However, Sodra will work with designers to help realise products.

Unit price vs capital investment
The cost of a press-forming tool is in the region of 35,000 euros.

Speed
The components have to be left in the press for approximately seven minutes and a fair amount of handwork is required to finish them.

Surface
As expected, the surface looks and feels like ultra-stiff cardboard. Colours, approved for use in children's products, are impregnated within the paper before forming takes place.

Types/complexity of shape
Shapes are based on a standard male and female configuration, so undercuts are not feasible.

Scale
At the time of writing the maximum size of the press tool is 500 × 1,000 millimetres.

Tolerances
Thickness dimensions are in the range of 1.8 millimetres to 10 millimetres (although 10 millimetres has yet to be trialled).

Relevant materials
DuraPulp is made from paper pulp mixed with PLA.

Typical products
Products currently being trialled are in the area of furnishings, such as chairs and lighting.

Similar methods
Moulding paper pulp (p.149), paper-based rapid prototyping (p.242).

Sustainability issues
DuraPulp is made using paper pulp and organic plastic both of which are non-toxic and renewable. At the end of its lifespan the material will biodegrade without harming the soil, avoiding landfill disposal. However, the process involves several manufacturing stages and is quite heat intensive, so has high energy use.

Further information
www.sodrapulplabs.com

+

- **An alternative to plywood.**
- **Economical material usage.**

–

- **Process is not yet fully commercialised.**

Bending Plywood

Product	AP Stool
Designer	Shin Azumi
Materials	plywood
Manufacturer	Lapalma srl
Country	Italy
Date	2010

This elegant, stackable stool is formed from a single sheet of plywood, with the seat and the body of the stool merging seamlessly. The wide spread of the base helps to disperse pressure.

The conversion of a tree into a simple-looking piece of bent plywood furniture involves at least 35 steps. The technique of cross-laminating veneers to produce stable, stiff engineered materials was first understood by the ancient Egyptians, who used the process for making items such as their iconic sarcophagi. The development of modern bent plywood is the result of a range of technological advancements, including the ability to cut the veneers accurately, the presses to laminate them and the glues to construct them.

The processing of most natural materials tends to be concentrated in the locations where the materials originate, and plywood production takes place mainly in northern Europe, North America, Southeast Asia and Japan. Starting from the point where the veneers have been sliced or rotary-cut from the logs, these large strips are cut into individual sheets, which are subsequently dried by being passed through a long chamber, at the end of which they are stacked according to quality.

The veneers are fed into rollers that distribute an even layer of glue over each sheet, with the quantity of glue being determined by the porosity of the timber. The various sheets are then stacked, with the grain running in alternate directions, to form an odd number of layers. The assembled sheets are placed over the female part of a mould, with the male part clamped on

top. The moulds allow for an excess of veneers, which is trimmed to form neat edges once the glue has dried. Depending on the shape, a pressure of several tonnes is needed to compact the sandwich together. The vertical pressure is aided by horizontal pressure, forcing the moulds to come together from all sides, and a combination of heat and pressure cures the glue. The part stays in the mould for about 25 minutes, the exact time depending on the shape. In industrial production, a CNC (computer numerical controlled) cutter is then used to trim the uneven layers to form a clean edge.

Volumes of production
Jigs can be made up for single profiles in a small workshop. Industrial production set-ups can produce hundreds of thousands of units.

Unit price vs capital investment
Jigs for low-volume production can be expensive due to the labour costs involved. However, depending on the design, simple moulds can be made that are still economical for small production runs or even one-offs. At the industrial end of the scale, as with most other manufacturing processes, higher tooling costs are balanced by low unit costs.

Speed
Cycle times are fairly long because the glued veneers have to dry inside the mould before they can be taken out, and the parts need subsequent finishing, including edge trimming, surface treatments or painting.

Surface
Dependent on the type of wood.

Types/complexity of shape
Restricted to simple bends in a single direction. The inherent flexibility of the material can allow for slight undercuts when removing pieces from the moulds.

Scale
The scale is generally suited to furniture and accessories (such as magazine racks). The restriction on size is determined by the size of the moulds and the ability to exert the degree of pressure needed to compact the layers.

Tolerances
Rather low because of the flexibility of the material.

Relevant materials
Birch is used in the majority of mass-produced furniture, but many other types of timber, including oak and maple, can be used. Burly woods (the outgrowth of a tree, also known as the burr), including pine, are not to be recommended, because it is difficult to produce plywood of consistent quality from them.

Typical products
Furniture, interiors and architectural cladding. Aircraft frames were made from bent plywood during the two World Wars.

Similar methods
Inflating wood (p.184) and pressing plywood (p.86).

Sustainability issues
Wood is a natural resource that is renewable when controlled by a sustainable forestry. However, the production of the plywood, and additional forming, is quite energy intensive as the wood undergoes extensive processing. At the end of its lifespan the plywood can be recycled.

Further information
www.woodweb.com
www.woodforgood.com
www.artek.fi
www.vitra.com
www.lapalma.it

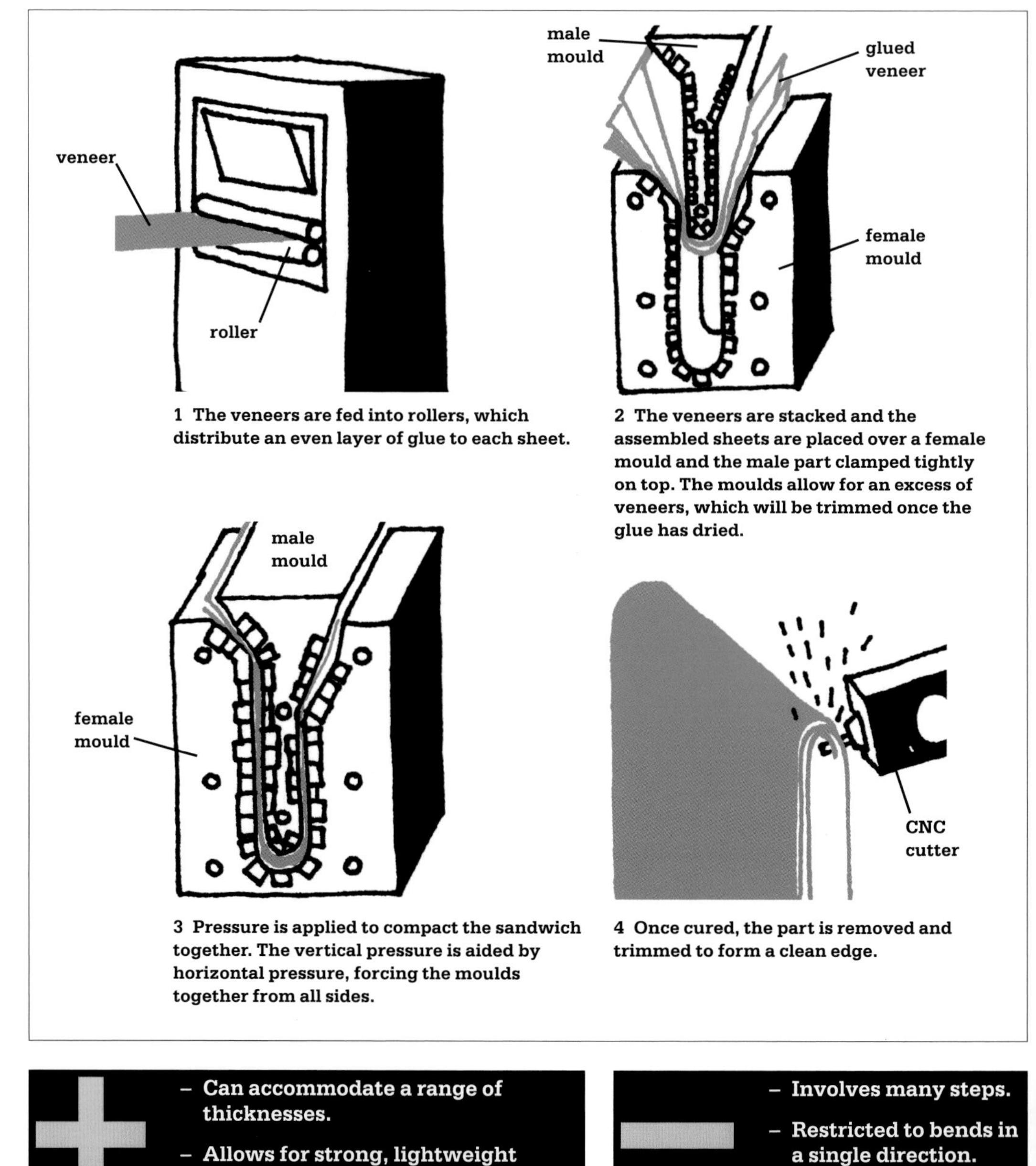

1 The veneers are fed into rollers, which distribute an even layer of glue to each sheet.

2 The veneers are stacked and the assembled sheets are placed over a female mould and the male part clamped tightly on top. The moulds allow for an excess of veneers, which will be trimmed once the glue has dried.

3 Pressure is applied to compact the sandwich together. The vertical pressure is aided by horizontal pressure, forcing the moulds together from all sides.

4 Once cured, the part is removed and trimmed to form a clean edge.

+

- Can accommodate a range of thicknesses.
- Allows for strong, lightweight components.

–

- Involves many steps.
- Restricted to bends in a single direction.

Deep Three-Dimensional Forming in Plywood

This book features a number of new and radical ways to process plywood into ever more complicated and curved forms. One of these is deep three-dimensional forming in plywood, a combination of a production method and a material specifically developed for the purpose. Using an innovative treatment in which the wood fibres are relaxed, it is now possible to bend plywood into wavy shapes that were once unthinkable.

The technology for preparing the plywood was developed by the German manufacturer Reholz®, and it enables the plywood to be moulded into a deep, three-dimensional compound curve, which is capable of producing forms that resemble moulded plastic rather than a piece of wood.

The key to this process, and the first stage in achieving the complex curves, is a series of closely cut, parallel lines, so deeply cut into the individual veneers that the wood itself almost falls apart. This gives the veneers the elasticity to be bent in different directions without breaking, which is particularly important when bending them against the direction of the grain. The individual sheets of veneer are glued together in a way that is similar to the bent wood process (see p.80), in order to achieve stiffness and strength.

Product	Gubi chair
Designer	Komplot
Materials	Walnut veneer
Manufacturer	Gubi using Reholz® deep 3D forming technology
Country	Germany (process) Denmark (chair)
Date	2003

The apparently simple curves of the Gubi chair mask the sophistication of this completely new method of forming wood. The compound curves (the seat passes through an almost 90 degree angle) are a result of the ability of this treated ply to conform to far more complex shapes than, for example, the pressed plywood tray (see p.86).

Volumes of production
Suitable for mass-production.

Unit price vs capital investment
The process is not cheap, but it is the only way to achieve this level of freedom in shaping plywood, which can go a long way in justifying the investment in tooling for large production runs.

Speed
The Reholz® process involves several steps, including the pre-treatment of the veneer, pressing and, finally, trimming. The real difference, however, between standard bent plywood manufacturing (see p.80) and the deep three-dimensional forming process is just one step – the cutting treatment to which the basic veneers are subjected that enables the plywood to be bent.

Surface
Comparable to any other type of close-grained wood surface, which can be stained, painted and coated in a number of ways.

Types/complexity of shape
The key here is that deep three-dimensional forming allows layers of thin, prepared veneer to be built up into formed sheets of plywood capable of being bent into previously unachievable curves. Designs should allow for the component to be separated from male and female moulds, so there can be no undercuts.

Scale
The scale is limited by the size of the veneers available and the moulds used to form the final shape.

Tolerances
Because of natural variations in the grain of the wood, no moulding is exactly like any other and it can be difficult to achieve very high tolerances. This can, nevertheless, be dealt with in a number of ways, for instance by using flexible fixings.

Relevant materials
Although many basic veneers can be used for deep three-dimensional forming, it is important that the veneer should have a straight grain direction and no knots. High-quality ALPI-veneers are especially suitable (these, recommended by Reholz® who developed the technology, are veneers produced by the Italian company, ALPI).

Typical products
Chair seats, plywood bars, bent furniture frames or, on a larger scale, laminated wood structures for use in the construction industry. Deep three-dimensional formed plywood can also be used to coat the housings of medical devices, for example, panelling for MRI (magnetic resonance imaging) scanning machines and in packaging to replace MDF (medium density fibreboard), mouldings for lights and parts of automobile interiors.

Similar methods
The manufacturer compares this process to the deep-drawing of metal sheets. However, in terms of wood production the closest method is bending plywood (p.80), although this has the disadvantage of being formable in only a single direction. The inflated wood process (p.184) developed by Malcolm Jordan allows similar three-dimensional forms to be produced, but it requires both foam and plywood. Pressing plywood (p.86), which is used to make dinner trays and car dashboards, achieves a similar three-dimensional effect, but with much shallower results.

Sustainability issues
Wood is a natural resource that is renewable when overseen by sustainable forestry. However, production is quite energy intensive as several processes are involved in treating and pressing the veneer.

Further information
www.reholz.de

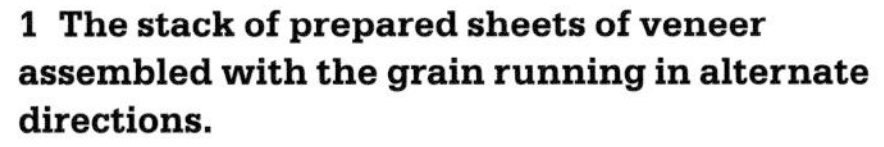

1 The stack of prepared sheets of veneer assembled with the grain running in alternate directions.

2 The stack of veneers before the male and female moulds are brought together.

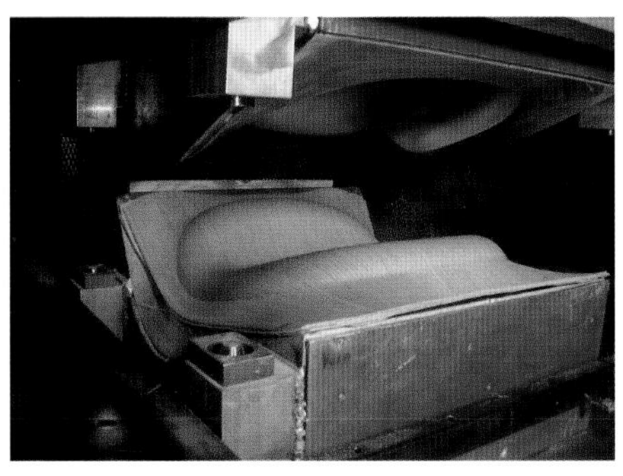

3 The seat for the Gubi chair, post forming.

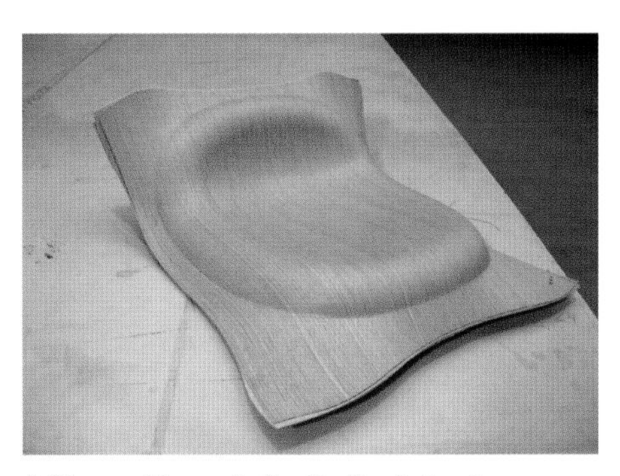

4 The seat is ready for the final shaping process, which involves the excess material around the seat being cut away.

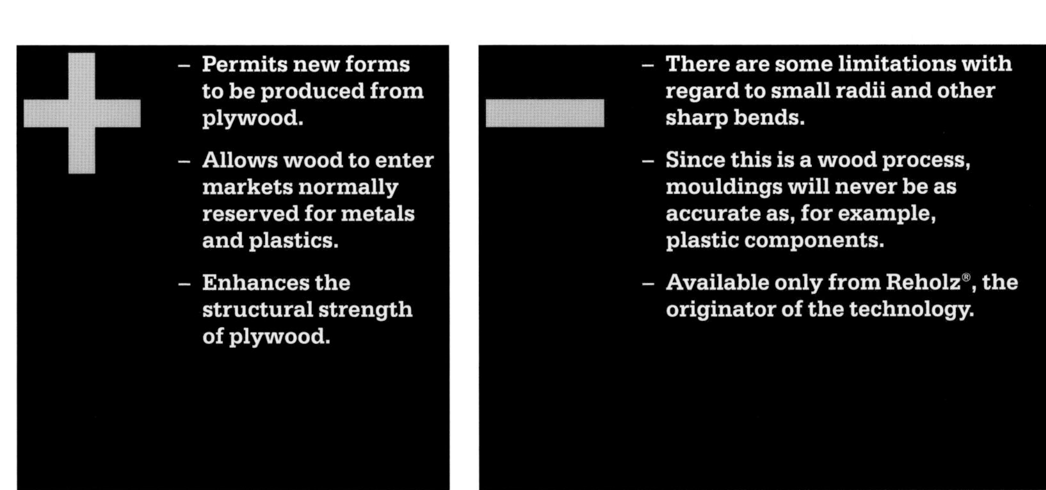

+

- Permits new forms to be produced from plywood.
- Allows wood to enter markets normally reserved for metals and plastics.
- Enhances the structural strength of plywood.

–

- There are some limitations with regard to small radii and other sharp bends.
- Since this is a wood process, mouldings will never be as accurate as, for example, plastic components.
- Available only from Reholz®, the originator of the technology.

Pressing Plywood

The first noteworthy thing about this production method is that the products it is used for are formed by a method that is more reminiscent of plastic forming than wood forming. By that I mean that they are formed from a flat sheet of wood into a three-dimensional shape in a way that gives results similar to shallow plastic vacuum thermoforming (see p.64).

The first stage of the process for making the archetypal trays you find in canteens around the world involves the raw material of veneers being cut and trimmed into square sheets. In most cases a single layer of veneer will be made up of two narrow leaves which are 'sewn' together with a flat, cross-hatched thread of glue. The sheets are stacked together and arranged with the grain running in alternate directions and with sheets of glue-impregnated paper in between each sheet. A melamine-impregnated sheet is added to the top and bottom of this stack, like the bread in a sandwich. The sandwiched packs are then placed in a press, in between male and female moulds, where pressure is applied for approximately four minutes at 135°C. It is important for the veneers to have a good degree of moisture to prevent the wood from splitting. Once removed from the press, the trays are stored on a flat table and held down with weights to ensure that they do not warp. The final stage involves the edges being trimmed and sealed with a spray of a clear lacquer.

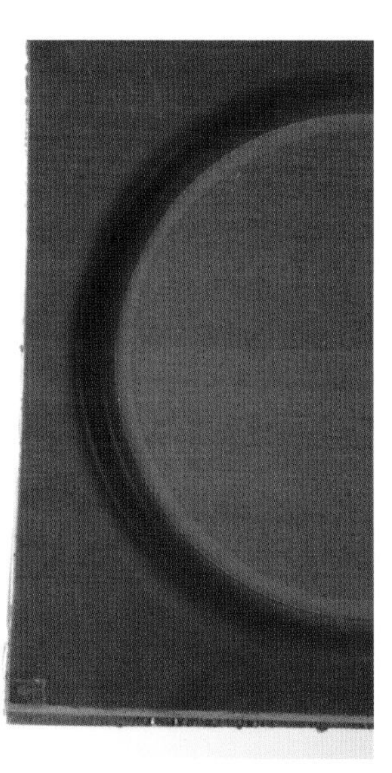

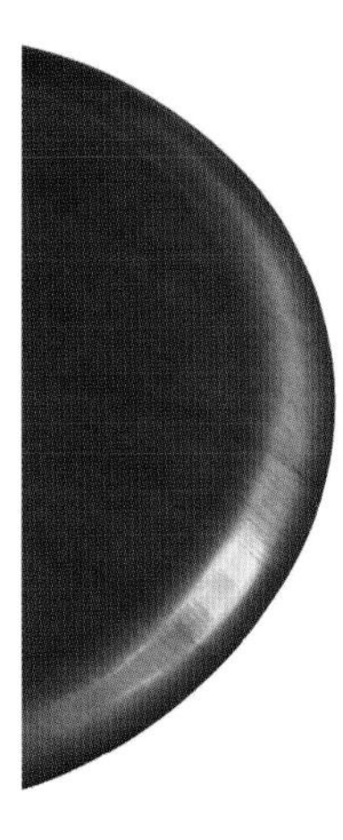

Product	**dinner tray**
Designer	not applicable
Materials	lacquered birch
Manufacturer	Neville & Sons
Country	UK

This sequence of images shows the layers of veneers and glue-impregnated paper; a pressed, untrimmed plywood tray; and the trimmed, lacquer-sealed tray.

The combination of heat, pressure and adhesive enables a range of laminated wood products to be produced, resulting in thin-section designs that can be extremely strong. Neville and Sons in the UK have been making an assortment of trays for several years. Today, they are one of the few remaining UK-based companies still producing wooden laminated trays. It is through the action of heat and pressure that they can make durable trays of only about 15 millimetres deep.

Volumes of production
As many as 600 wooden trays can be produced in a day. Minimum orders from Neville and Sons are 50 units.

Unit price vs capital investment
An affordable ratio for small-scale production makes this method of forming highly suitable for small- and large-scale runs. Moulds for trays are produced in aluminium covered in a stainless steel sheet, which makes them cost-effective even for batch production. Unit prices are very low.

Speed
One tray can be produced every five minutes.

Surface
The surface colour, and to a degree the finish and pattern, are controlled by the melamine sheet that is used in the pressing process. Decorative patterns, colours and non-slip surfaces are available.

Types/complexity of shape
Embossed sheets with a fairly low-draw impression, up to a maximum of approximately 25 millimetres.

Scale
Neville and Sons can produce products measuring up to 600 by 450 millimetres.

Tolerances
Not applicable.

Relevant materials
Most veneers are suitable. However, the material used for trays is generally birch, beech or mahogany.

Typical products
Given the shallowness of the depth achievable, this process is limited to products such as trays and automotive trim with the kind of walnut effect you would expect to find in high-end brands.

Similar methods
A process that allows for much more depth and possibilities with curving plywood is deep 3D-forming (p.83). Also relevant, but using a completely different technology, is the inflated wood by Curvy Composites (p.184).

Sustainability issues
The cross-directional grain structure of plywood minimises material consumption while maintaining excellent strength. Pressing the plywood in the form of the tray featured here requires a number of processes, including a fair amount of heat to melt and bond the glues and veneers.

Further information
www.nevilleuk.com

+

- **Extremely durable: heat-resistant and dishwasher-safe.**
- **Excellent chemical resistance.**
- **Printable surface.**

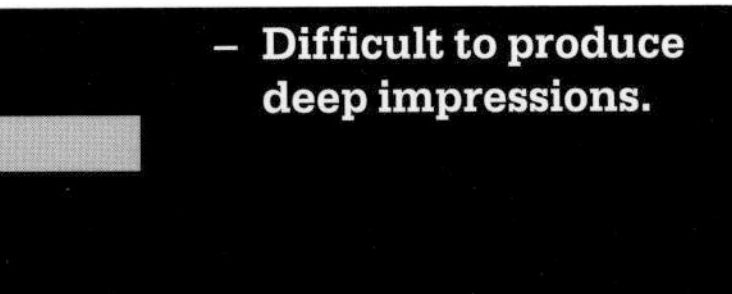

–

- **Difficult to produce deep impressions.**

Contin

uous

Components that are made from continuous lengths of a material

This chapter looks at components that are made according to the same principles that are used to make sausages, or, in other words, components that are the result of material being fed through a shape to produce long lengths of the same profile. It also looks at continuous strips of wood and plastic, woven lengths of metal and continuous lengths of bent steel. It celebrates a rich assortment of processes that use a range of dies to form materials that can be produced in infinitely long length but which, with one exception, have the same cross-sectional shape along the whole length. Many of these processes are extremely cost-effective because they can produce identical multiples, cut from the same strip or section.

Calendering

Calendering has traditionally been used as a finishing process applied to textiles and paper, using heat and pressure to give a smooth, shiny surface. In the nineteenth century, however, it was developed so that multiple rollers could produce rubber sheet. It is a large-scale process, both in terms of the volumes it can produce and the sheer size of the machine that is used to form the sheet material itself (or to add texture to existing sheet materials).

Imagine a machine that has at its heart a series of steel rollers resembling a clothes mangle, that press materials into continuous lengths of thin sheet. Although still used for finishing paper, some forms of textiles and various types of elastics, calendering is the preferred method for forming high volumes of PVC sheet at a fast rate. In the realm of plastic production it competes with extrusion (see p.96) in the production of both rigid and flexible plastic sheet.

When it is used in this sort of plastic production, the set-up usually includes at least four heated rollers, rotating at different speeds. Before this, however, hot granules of the plastic are fed into a kneader where they reach a gelling stage. They are then fed, via a conveyor belt, through the first of the heated rollers. The rollers are carefully controlled to produce the correct thickness and finish. Embossing rollers can be used to add texture and the sheet material then passes through cooling rollers prior to being wound onto a giant roll.

The essence of calendering is captured in this image, which shows a ribbon of plastic being passed over polished steel rollers to form a continuous length of plastic sheet.

Volumes of production
Because of the set-up costs and running times involved, tooling is expensive, making calendering exclusively a very high-volume production process. The minimum length for production varies between 2,000 and 5,000 metres, depending on the gauge of the sheet.

Unit price vs capital investment
Calendered sheets are often further converted in order to be turned into products, but the price of the sheet before it is converted is highly cost-effective if large enough orders are fulfilled. Capital investment is extremely high.

Speed
Once the process is running at optimum speed – which can take several hours to achieve – it is super-fast.

Surface
The rollers can be ultra-smooth to give a shiny surface, or embossed with patterns that are transferred onto the final sheet.

Types/complexity of shape
Flat, thin sheet.

Scale
The thickness of a PVC sheet is generally between 0.06 and 1.2 millimetres. The width of the rolls is up to 1,500 millimetres.

Tolerances
Not applicable.

Relevant materials
Calendering can be used for a range of materials, including textiles, composites, plastics (mainly PVC) or paper, where it is used to smooth the surface.

Typical products
Paper, used for newsprint, and large-scale plastic sheet or film. It can also be used as a finishing process for other papers and textiles.

Similar methods
In terms of plastic production, extrusion (p.96) is the closest comparable method for producing continuous sheet material. There is also blown film (p.92).

Sustainability issues
Calendering is a largely automated process that requires continuous heat and rotation and so relies on large amounts of power. However, the machines operate at incredibly high speeds to reduce cycle times which maximises this energy use. Excess material is minimal.

Further information
www.vinyl.org
www.ecvm.org
www.ipaper.com
www.coruba.co.uk

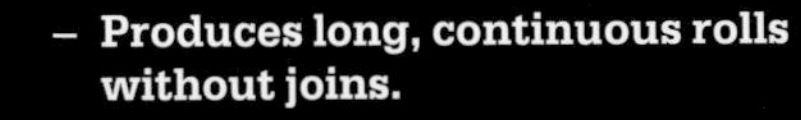

+
- **Produces long, continuous rolls without joins.**
- **Excellent method for producing large quantities of flat sheets.**

–
- **Suited to very large-scale production only.**

Blown Film

The best way to summarise the blown-film process is to think of blowing bubblegum, but on a giant industrial scale. Producing plastic that is on the physical scale of a building involves a massive tubular bubble of inflated plastic being blown upwards into a vertical scaffolding structure.

The technique takes its name from the action of the plastic granules, which are heated (1) and fed vertically by a stream of air through a horizontally placed cylindrical die (2), to form a thin-walled tube that is blown to form a huge plastic bubble (3). This bubble is fed vertically by a stream of air in the top of the die to form a tower of plastic (4). Varying the volume of air in the bubble controls the thickness and width of the film,

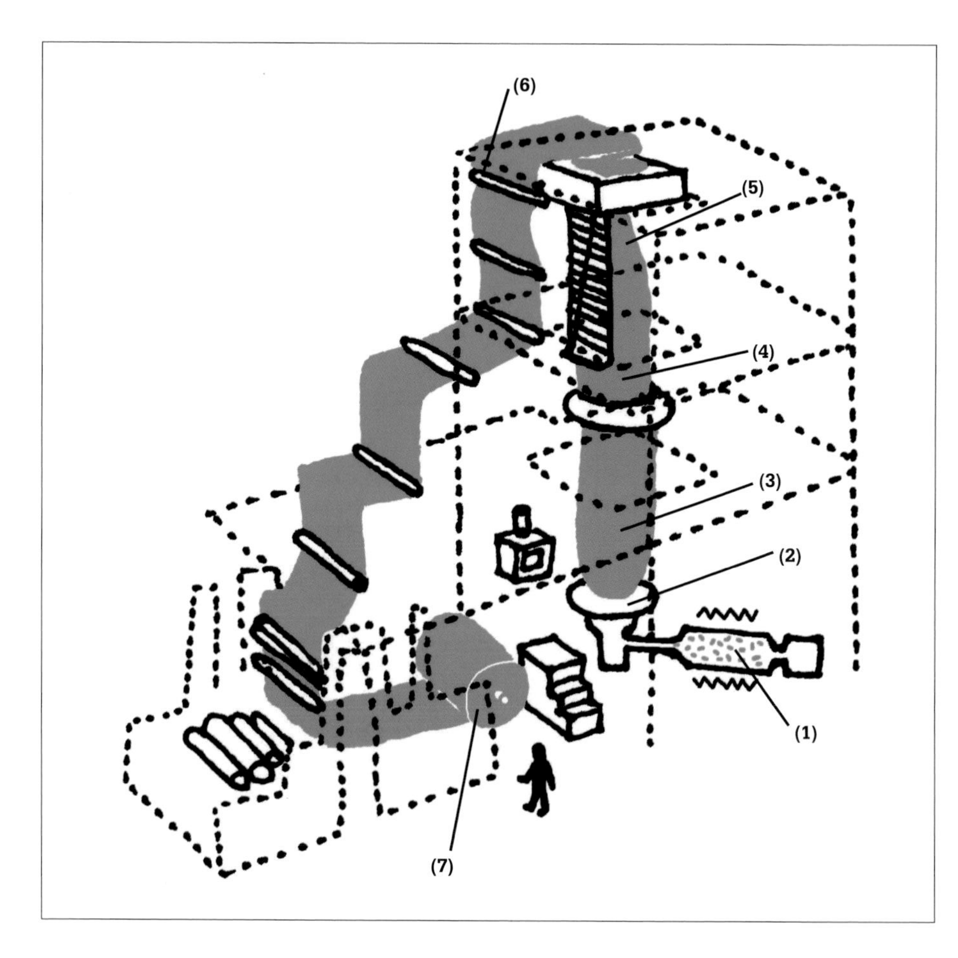

which gradually cools as it rises, tapers and, several metres up, eventually subsides completely into a flattened tube (5). This flat tube passes through a series of rollers on its way back down to ground level (6), where it is wound onto a giant roll, ready for despatch (7). The edges of this sheet material can be trimmed off to produce sheets, or it can be left as a tube to be used for supermarket bags and bin-liners.

Volumes of production
This is a high-volume method with a capacity to convert 250 kilograms of plastic per hour.

Unit price vs capital investment
High capital costs, but extremely cost-effective for large production runs.

Speed
Up to 130 metres per minute.

Surface
Controlled by various factors, including the material and machinery set-up.

Types/complexity of shape
Flat sheets or tubes only.

Scale
Blown films range from 550 millimetres to 5 metres in diameter, and can be up to hundreds of metres in length. Films are available in thicknesses of from 10 or 20 microns up to 250 microns.

Tolerances
This process can achieve high tolerances, but you should be aware that some manufacturers offer two grades of blown film – with and without thickness control.

Relevant materials
The most common materials are high- and low-density polyethylene, but other materials, such as polypropylene and nylon, can also be used.

Typical products
Most plastic film products, such as bin-liners, carrier bags, sheeting, cling film, laminating film and just about any other type of film you care to mention.

Similar methods
Extrusion (p.96) and calendering (p.90) are both used to produce thin flat sheets.

Sustainability issues
The blown film process is capable of producing incredibly high volumes of plastic film at a very fast rate for minimal cycle times and therefore minimal energy consumption. However, energy use is still substantial due to the high temperatures involved and the constant high pressures that are required for working. Excess waste material is minimal.

Further information
www.plasticbag.com
www.flexpack.org
www.reifenhauser.com

+

- **Allows for the production of a material with uniform properties across the whole length and width.**

- **Blown film is not always ideal – for example, the process of casting film can be a better option for applications that require high optical clarity.**

Exjection®

Injection moulding and extrusion are both vital manufacturing methods for numerous plastic and metal products. However, there are certain limitations to the types of product each can be used to produce. For instance, extrusion allows for long, thin parts to be created but the entire length of the strip must have a single profile shape, which means variations along the piece cannot be produced. In contrast, complex shaped pieces such as ledges or closures can be created with injection moulding, but only in short lengths as the resin will not flow throughout long stretches without evidence of sinking in the final component. Exjection® brings together the benefits of both processes in components that are not only long but have variations and detailing along the length of the part.

The process relies on much the same principles as conventional injection moulding where molten plastic is forced into a mould. However, the addition of a movable cavity makes the process a little more interesting. As the molten plastic is injected into the mould it begins to fill the hollow space. As this space begins to fill, the end of the cavity starts to move horizontally along the full length of the mould at the same speed at which the plastic is being injected. As the cavity moves, more of the mould is exposed for the molten plastic to flow into. With the cavity moving and the plastic being forced in at the same speed, a high pressure is maintained which enables the consistent flow of the plastic without shrinkage. It might help to imagine filling a syringe. As you pull the back of the plunger, the pressure pulls the liquid into the container, and the more you pull back, the more area there is for the liquid to fill. The absence of air creates a vacuum which is under high pressure.

When the mould is completely full, the molten plastic is left to cool and set before it is removed and the cavity set back to its starting position for the next cycle. Material type and wall thicknesses have an impact on the speed of production as they affect how quickly the cavity can move.

Product	An Exjection® sample
Materials	POM (although Exjection® can be applied to a range of plastics)
Manufacturer	Exjection®
Country	Germany
Date	Exjection® was first presented in 2007

Sample of an Exjection® moulding that shows the small supports that cut cross the length. Conventional extrusion does not allow for this type of detail over a continuous shape.

Volumes of production
Comparable to injection moulding.

Unit price vs capital investment
The per unit cost of Exjection® moulding is comparable to that of a traditional injection moulding. However, the initial investment costs are higher because of the movable cavity.

Speed
The type of material and the wall thickness of the part define the speed of the cavity's movement and thus the cycle time.

Surface
Comparable to the high surface finish that is available from conventional injection moulding.

Types/complexity of shape
Exjection® allows for multi-cavity components, which means combined components can be manufactured in one cycle. It also allows for overmoulding.

Scale
Can vary widely.

Tolerances
± 0.1 millimetre.

Relevant materials
A wide range of commodity and engineering thermoplastics have been successfully processed. The overmoulding of metal and wood is also a possibility.

Typical products
The process is ideally suited to producing long, thin-walled parts, such as LED lighting strips, lamp covers and cable ducts with integrated mouldings such as closures and caps.

Similar methods
Exjection® is a proprietary process and there are no comparable methods.

Sustainability
Exjection® allows for several processes to be combined in one single cycle, which eliminates the need to transport parts between manufacturers, and so decreases emissions in addition to reducing lead time and therefore energy use.

Further information
www.exjection.com

+

- **Allows continuous profiles to be produced on an injection-moulding machine.**
- **Can cost less than comparable alternatives.**

–

- **Limited to a small number of manufacturers.**

Extrusion

Extrusion occurs in a variety of forms, from the low-tech squeeze of a toothpaste tube and the making of foods such as long-stranded pasta, to aluminum window frames and the continuous lengths of hard-boiled egg that McDonald's slices into its salads. In the simplest terms, extrusion is about squeezing a material through a hole in a die and producing continuous lengths of material at whatever profile that hole has.

This bench, or chair depending on the length you cut it, designed by Thomas Heatherwick shows a very large-scale example of extrusion. The project follows the simple premise of how can you make a seat, legs and back rest from a single shape in a single materials and eliminate any fixings or additional components. Heatherwick sought the largest extrusion machine in the world in order produce his vision. The piece also required a lot of polishing in order to transform the dull aluminium into a mirror-bright surface.

The piece captures the essence of the continuous nature of the production process, showing the unfinished tail that twists into space.

Close-up of extrusion die.

Detail from extrusion emerging from die.

Product	Extrusions
Designer	Thomas Heatherwick
Materials	aluminium
Manufacturer	Haunch of Venison
Country	UK
Date	2009

Apart from the large scale of this piece, the most wonderful aspect is the reminder of the extrusion process in the tail. It illustrates an aspect of extrusion that is always cleaned up and never seen.

Volumes of production
Different manufacturers have different minimum lengths, but extrusion can be a cost-effective process for both batch and large-scale runs. It is definitely not for one-offs – unless your one-off is 50 metres long.

Unit price vs capital investment
Conventional extrusion requires a low investment in tooling when compared with injection moulding (see p.196), for example.

Speed
Up to 20 metres per hour.

Surface
Excellent.

Types/complexity of shape
No problems in making complex shapes with varying wall thicknesses, just as long as the shape is the same along the whole length. Flat sheet can also be produced.

Scale
Depends on the type of extrusion. Most manufacturers have an average maximum size of 250 millimetres in cross-section. The length is limited by the size of the factory.

Tolerances
Difficult to maintain high tolerances due to a wearing of the die.

Relevant materials
Extrusion is a versatile process and can be used for materials including wood-based plastic composites, aluminium (shown in this example), magnesium, copper, and a wide variety of plastics and ceramics.

Typical products
Everything from architectural and furniture components, lighting and accessories, to pasta, and sticks of rock candy with place names written through them.

Similar methods
Pultrusion (p.99), calendering (p.90), coextrusion (multiple layers of extruded material in the same component), laminating (two, or more, materials bonded together), roll forming (p.104) and impact extrusion (p.146).

Sustainability Issues
There are multiple forms of extrusion, but both hot and cold extrusion require either high temperatures or pressure, which can be energy intensive. Extrusions can crack internally during forming when too much heat or pressure is applied, so parts must be monitored closely to prevent waste of materials. The nature of the long lengths of extrusions also means parts need post-cutting in order to convert them into usable shapes.

Further information
www.heatherwick.com
www.aec.org

+

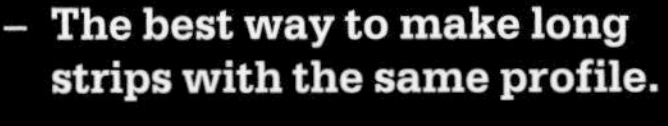

- **The best way to make long strips with the same profile.**
- **Can be used for a range of materials.**
- **Extensive production base.**

–

- **Parts often need to be cut to length, assembled or drilled.**

Pultrusion

Pultrusion is much less common as a plastic-processing method than its more familiar relative, extrusion (see p.96). The processes are similar in that they allow continuous lengths of a set and an unchanging profile to be formed, but one of the main differences between them is that extrusion can be used for aluminiums, wood-based composites and thermoplastics, while pultrusion is used in the forming of composites that use long strands of fibre as reinforcement.

As the name suggests, the process is based on pulling the blended materials of the composite through a heated die. This differs from extrusion, which is based on pushing the material. The continuous lengths of reinforcing fibres, which can be made from glass or carbon, are saturated with a liquid resin mixture as they are pulled through the die, which, besides shaping the component, also acts to cure the resin as it is heated. Sometimes, pre-impregnated ('pre-preg') fibres are used, removing the need for a resin bath.

Manufacturers of plastics have, in recent years, experimented with many applications that traditionally used metals, and pultrusion is a typical example of the benefits such experimentation can bring. Pultruded plastics display an increased range of physical properties that can benefit both engineering and design applications, because they offer the toughness of metals with the advantages of low weight and corrosion-resistance. Pultrusions are incredibly dense, hard and rigid sections – they even 'clank' like pieces of metal when you knock them!

Product	**sample of pultruded composite profile**
Materials	glass fibre and polyester resin composite
Manufacturer	Exel Composites
Country	UK

These profiles illustrate two of the key properties of pultrusion: first, its ability to produce shapes in plastic having similar properties to metal profiles; secondly, its capacity to have moulded-in colours.

Volumes of production
Depends on the size and complexity of the shape. 500 metres is a typical minimum run.

Unit price vs capital investment
The cost is lower than that for some moulding processes, injection (see p.196) and compression moulding (see p.174), for example, but higher than for, say, hand lay-up moulding (see p.152).

Speed
Depends on size, but, as a rule of thumb, it is possible to achieve 0.5 metre per minute for a profile measuring 50 by 50 millimetres, 0.1 metre per minute for chunky shapes and 1 metre per minute for narrower sections.

Surface
The surface finish can be controlled to a degree, depending on the reinforcement and polymer.

Types/complexity of shape
There are no problems with undercuts in pultrusion. Virtually any type of shape that can be squeezed through the die can be made, bearing in mind that the shape must have a constant thickness.

Scale
The maximum size for profiles is typically 1.2 metres wide, although there are specialist machines that make larger components. Minimum wall thickness is approximately 2.3 millimetres. The size of the manufacturing plant dictates the limit to the length of the pultrusion.

Tolerances
Vary depending on the profile, but on a standard box-section, with a wall thickness of 4.99 millimetres, the tolerance is ±0.35 millimetres.

Relevant materials
Any thermoset polymer matrix that can be used with glass and carbon fibre.

Typical products
Applications for pultrusions are varied and include permanent and temporary structural components for industrial plants, vandal-resistant indoor and outdoor public furniture, and funfair and exhibition stands. Smaller-scale applications include electrically insulated ladders, ski poles, racquet handles, fishing rods and bicycle frames. Perhaps surprisingly, pultruded plastics have a resonance similar to certain woods, which has led to them being used as replacements for hardwood frames for xylophones.

Similar methods
Extrusion (p.96) and Pulshaping™ (p.102).

Sustainability issues
Parts can be produced with thin wall thicknesses as a result of the fibre reinforcement, which minimises material use without compromising strength. However, as the process is entirely automated and heat intensive, energy use can be quite high in relation to the fairly slow cycle speed. The combination of materials makes the composites non-recyclable.

Further information
www.exelcomposites.com
www.acmanet.org/pic
www.pultruders.com

1 Individual strands of fibre are fed into a die where they will be soaked in resin and formed into their final profile.

2 A finished tube emerges through the cutter, ready to be cut to length.

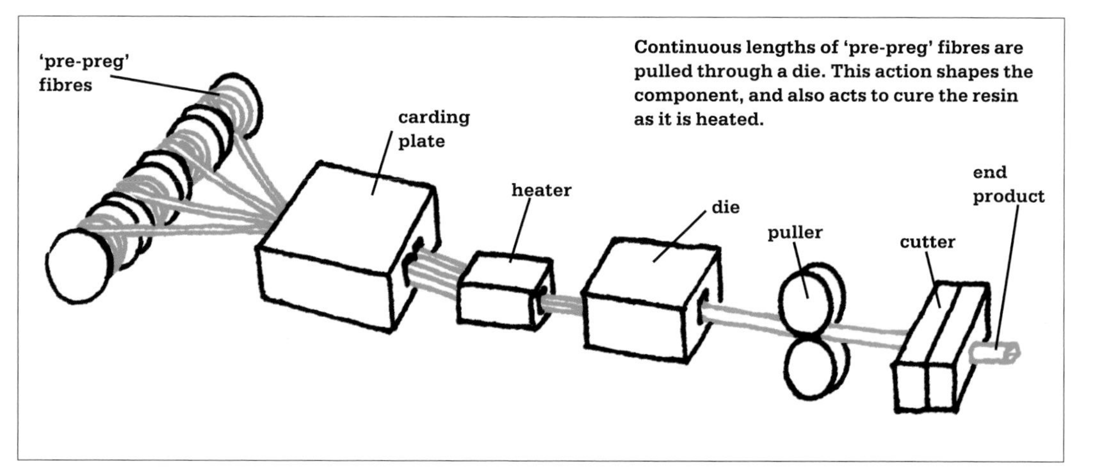

Continuous lengths of 'pre-preg' fibres are pulled through a die. This action shapes the component, and also acts to cure the resin as it is heated.

- Offers a 75 to 80 per cent weight reduction on steel and 30 per cent on aluminium.
- Greater dimensional stability than its metal counterparts.
- Can be coloured without the problem of chipping because the colour is added to the polymer itself.
- Surface decorations can be applied to mimic grain and other textures.
- Non-conductive and non-corrosive.

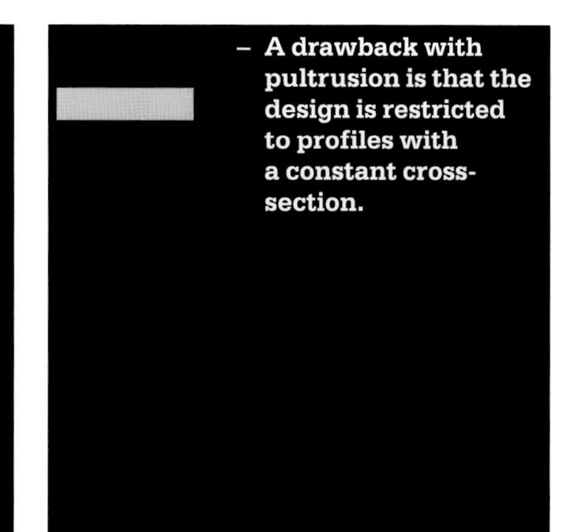

- A drawback with pultrusion is that the design is restricted to profiles with a constant cross-section.

Pulshaping™

Pulshaping™ is one of the newest additions to the manufacturing world and the processing of composites. Developed by US-based Pultrusion Dynamics, Inc., it addresses one of the biggest problems – the constant, unvarying cross-section along the whole length – in the pultrusion process (see p.99). Pulshaping™ allows designers to modify a cross-sectional shape in three dimensions during continuous processing of components in fibre-reinforced plastics. For example, a round profile can be a constant cross-section along the majority of its length, and can then be transformed to a square at one end and an oval at the other, using appropriate tooling. A particular advantage of this process is that it can, for example, allow tube ends to be shaped with threaded fasteners or expansion–reduction couplings joints.

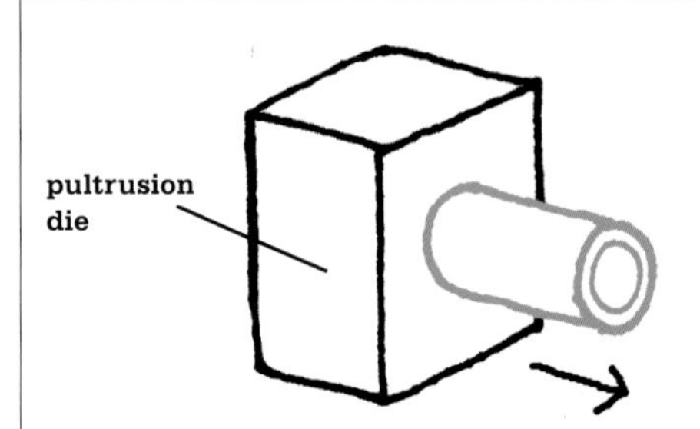

1 A standard pultrusion die is used to form a cylindrical cross-section.

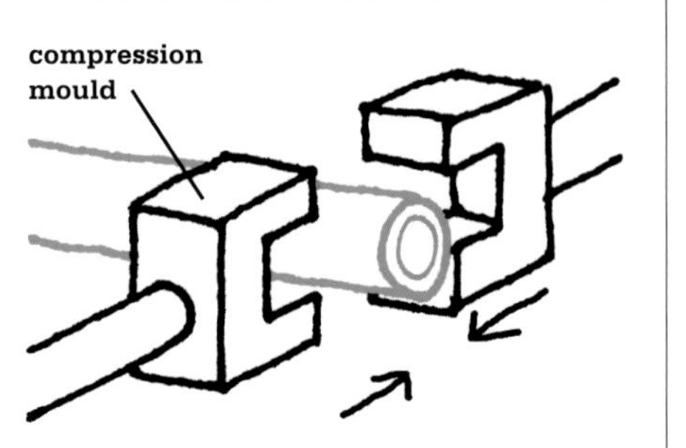

2 A two-part compression mould is used to apply pressure and thereby squash the cylindrical walls.

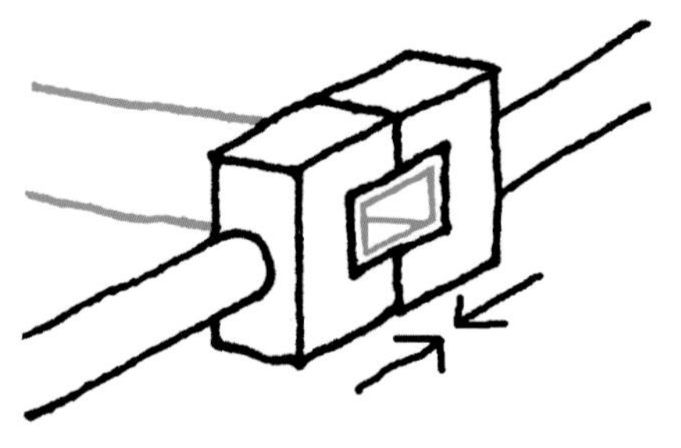

3 The pressure forms the tube into the desired cross-section.

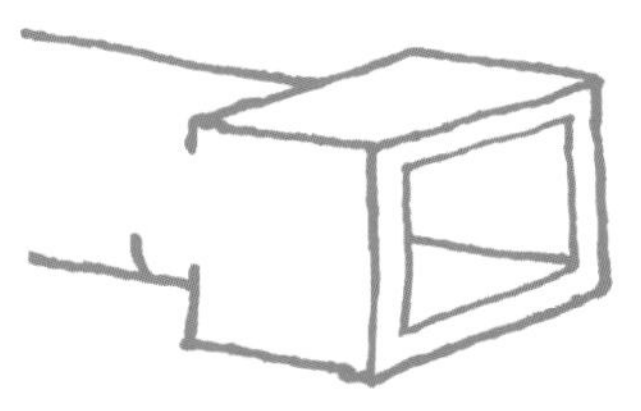

4 The finished part has been 'morphed' from the cylindrical to the new cross-section.

Volumes of production

Still in its developmental stage, Pulshaping™, like conventional pultrusion (see p.99), is potentially a high-volume production process.

Unit price vs capital investment

Pulshaping™ is a fairly costly process and is not suited to small production runs. Economic quantities are in excess of 2,000 lineal metres.

Speed

Typically 0.5–1 metre per minute for the continuous pultrusion part of the process, plus an additional 1–3 minutes for reshaping cycles.

Surface

As in pultrusion, the surface finish can be controlled to a slight degree of variation, depending on the reinforcement and polymer. Due to the ability of the process to allow manipulation of form along the cross-section surface, features such as dimples and projections can be designed into the product in the reshaped segment.

Types/complexity of shape

This process is highly versatile because of its ability to produce a variety of cross-sectional shapes.

Scale

Ideally suited to long products, over 1.8 metres.

Tolerances

Very fine tolerances.

Relevant materials

Thermosetting resins with glass, carbon or aramid fibre.

Typical products

Components such as handles for large tools, which typically require the main body to be straight, with end features to be produced in separate processes, can be made in one go with Pulshaping™.

Similar methods

There is nothing really similar to this process in the sense that similar methods, such as extrusion (p.96) and pultrusion (p.99), do not allow for manipulation of the cross-section.

Sustainability issues

As with pultrusion, parts can be produced with thin wall-thicknesses as a result of fibre reinforcement, to minimise material use while optimising strength. The combination of materials makes the composite non-recyclable.

Further information

www.pultrusiondynamics.com

+

- **Shares the many advantages listed for pultrusion (see p.99).**
- **The added advantage is that the geometry can be changed at selected locations along the continuous length of a component.**

–

- **Although the geometry can be altered along the length of the component, this is restricted to a repeat pattern. A continuous curved shape or continuous taper cannot be executed with this method.**

Roll Forming

Roll forming can be used to produce continuous lengths of anything from simple shapes in a single operation to quite complex profiles that require a number of passes through different rollers, from square sections to round shapes and from folded flanges to box sections.

In simple terms, roll forming involves passing a continuous sheet of metal, plastic or even glass, over or through a series of at least two shaped rollers. Feeding the sheet in a straight line between the rollers forces the material to bend into the required profile. The bending occurs progressively over the series of rollers, in a process that may require up to about 25 different rollers, depending on the complexity of the profile. Roll forming can be achieved either as a cold forming process or with heat. In the case of glass, the sheet passes through the rollers as a molten ribbon.

1 A very crude set-up, but this shows a flat strip of metal fed into rollers to be bent into a fairly shallow radius.

2 As the distance between rollers is closed for this second pass through the rollers, so a curve with a tighter radius is achieved.

Product	**Apple iMac aluminium stand**
Designer	Apple Design Studio
Materials	aluminium
Date	2004

The aluminium stand for this iMac illustrates, in a discreet way, Apple's achievement in exercising extremely tight control over the manufacturing of their products. The achievement here is in being able to bend such a thick piece of aluminium without any tearing of the material at its widest radius, which would normally be associated with this thickness of material at this scale.

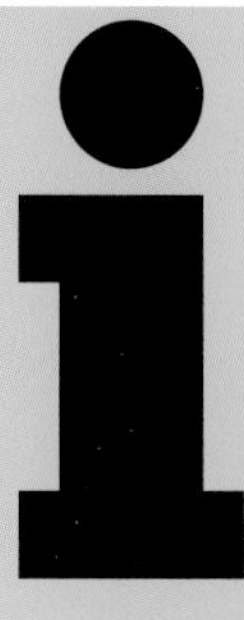

Volumes of production
High-volume mass-production.

Unit price vs capital investment
Set-up and tooling costs are high, which is why the process is suited to mass-production. It is, however, possible for small prototypes to be produced in a small workshop, depending on the complexity of the shape.

Speed
Production speeds are typically 300 to 600 metres per hour for a medium-sized manufacturer, depending on the complexity of the profile and the gauge of the material. Larger manufacturers can often go faster, but minimum quantities and lengths apply.

Surface
Other operations, such as punching and embossing, can be incorporated into the process to allow for surface details.

Types/complexity of shape
Long lengths of the same profile, which can be quite elaborate.

Scale
For mass-produced components the standard depth is approximately 100 millimetres, but it is possible to produce extremely large pieces, as demonstrated by the famous monumental curved steel structures by the artist Richard Serra. In theory, the only thing that dictates the length is the physical size of the manufacturing plant.

Tolerances
Vary between ±0.05 and ±1 millimetres, depending on the thickness of the sheet.

Relevant materials
Roll forming is almost exclusively used for forming metals, but it is also a useful process for glass and plastics, albeit on a much smaller scale.

Typical products
Car parts, architectural profiles, window and picture frames, and guides for sliding doors and curtain rails. In the case of glass, the process is employed to make U-shaped glass profiles that are used in architectural glazing.

Similar methods
For metal work, similar methods include sheet-metal forming (p.50) and extrusion (p.96), both of which also provide long lengths of a profiled shape.

Sustainability issues
Roll forming is a straightforward heatless process that produces little waste material and has a fairly quick cycle rate that minimises energy consumption. However, when using metals roll forming can in some cases cause microcracks and thinning, so sufficient testing should be carried out prior to production.

Further information
www.graphicmetal.com
www.crsauk.com
www.pma.org
www.britishmetalforming.com
www.steelsections.co.uk
www.corusgroup.com

- **Flexible in terms of finished length.**

- **Limited to an unvariable thickness of material.**

Rotary Swaging
AKA Radial Forming
with stationary-spindle and flat swaging

To explain this process in very simple terms, rotary swaging is used to alter the diameter of a range of metal tubing, rods and wires. The process involves the original material being fed through a series of rotating steel dies, which form the material to the required profile (which is always symmetrical and round). As they are rotating, the dies perform a hammering action at a rate of up to approximately 1,000 hits per minute, basically battering the work piece into shape.

Other forms of rotary swaging include stationary-spindle swaging, which is used to form non-round parts. Flat swaging is used to reduce the overall thickness of sheet metal.

The original-diameter material is fed into a rotating steel die (1). This hammers the material into shape with a series of backers (2), which hit the rollers as the piece rotates. The hammering takes place when the backers pass over a series of rollers (3). Simple centrifugal forces allow the backers to recede from the die before once again being pushed forward as they pass over the rollers.

Volumes of production
Medium to high levels of mass-production.

Unit price vs capital investment
Although the process sounds complicated, it is actually based around a very simple principle that involves minimal tooling and fast set-up times. This makes it unusual in that it is a high-volume process that is also cost-effective for short runs.

Speed
Simple shapes can be produced at a rate of 500 units per hour.

Surface
Rotary swaging gives an excellent, shiny surface as a result of the hammering, which acts to buff the surface. The finish is better than stock tubing that has not been swaged.

Types/complexity of shape
Because of the action of the rotating tool, options are limited to symmetrical and round shapes. All shapes of tubing, rod and wire can be converted into round profiles using this process, but stationary-spindle swaging needs to be used to obtain non-round sections.

Scale
Depending on the type of machinery available at the manufacturer, dimensions can vary from 0.5 up to 350 millimetres.

Tolerances
Good control of both the inside or outside diameter, depending on how the dies are set up.

Relevant materials
Ductile metals are the most commonly used. Ferrous metals with high carbon contents can be problematic.

Typical products
Golf clubs, exhaust pipes, screwdriver shanks, furniture legs and rifle barrels.

Similar methods
Machining (p.18), impact extrusion (p.146) and deep metal drawing (used to stretch a metal sheet into a variety of hollow shapes, such as cylinders, hemispheres and cups).

Sustainability issues
Rotary swaging is most commonly cold-worked so does not require heat, which means energy use can be significantly reduced. Additionally, there is no loss of material during manufacture and parts have increased strength after being worked, which improves durability and product life-expectancy.

Further information
www.torrington-machinery.com
www.felss.de
www.elmill.co.uk

+

- **A large range of symmetrical profiles can be formed.**
- **Because no metal is removed, the process is economical in its use of material.**
- **It is possible to achieve a fine degree of dimensional control of both the inside and outside surfaces.**
- **Working the material hardens it, thus increasing its strength.**

–

- **Rotary swaging is limited to forming round, symmetrical shapes (stationary-spindle swaging, however, can achieve non-round shapes, including squares and triangles).**
- **Reduction of diameter tends to be easier at the ends than at the middle of the tubing.**

Pre-Crimp Weaving

Pre-crimp weaving is a great case study in how unexpected materials can be woven and used decoratively. In the same way that soft fabrics are woven for decoration, rigid lengths of wire can be woven to dress and adorn our urban landscapes. Industrial weaving takes many forms, from the chain mail of industrial fencing and fabrics to architectural cladding. Although not recognised as a major industrial process, pre-crimp weaving can be utilised as a way to design large-scale decorative metal screens.

Product	**architectural mesh**
Materials	stainless steel and brass
Manufacturer	Potter & Soar
Country	UK
Date	2005

Architectural mesh can be produced to a wide range of specifications, to increase or decrease density, texture and transparency. Different optical effects can therefore be created and, in addition, it is self-supporting so it can be used for ceilings and cladding, as well as ornamental balustrading and furniture.

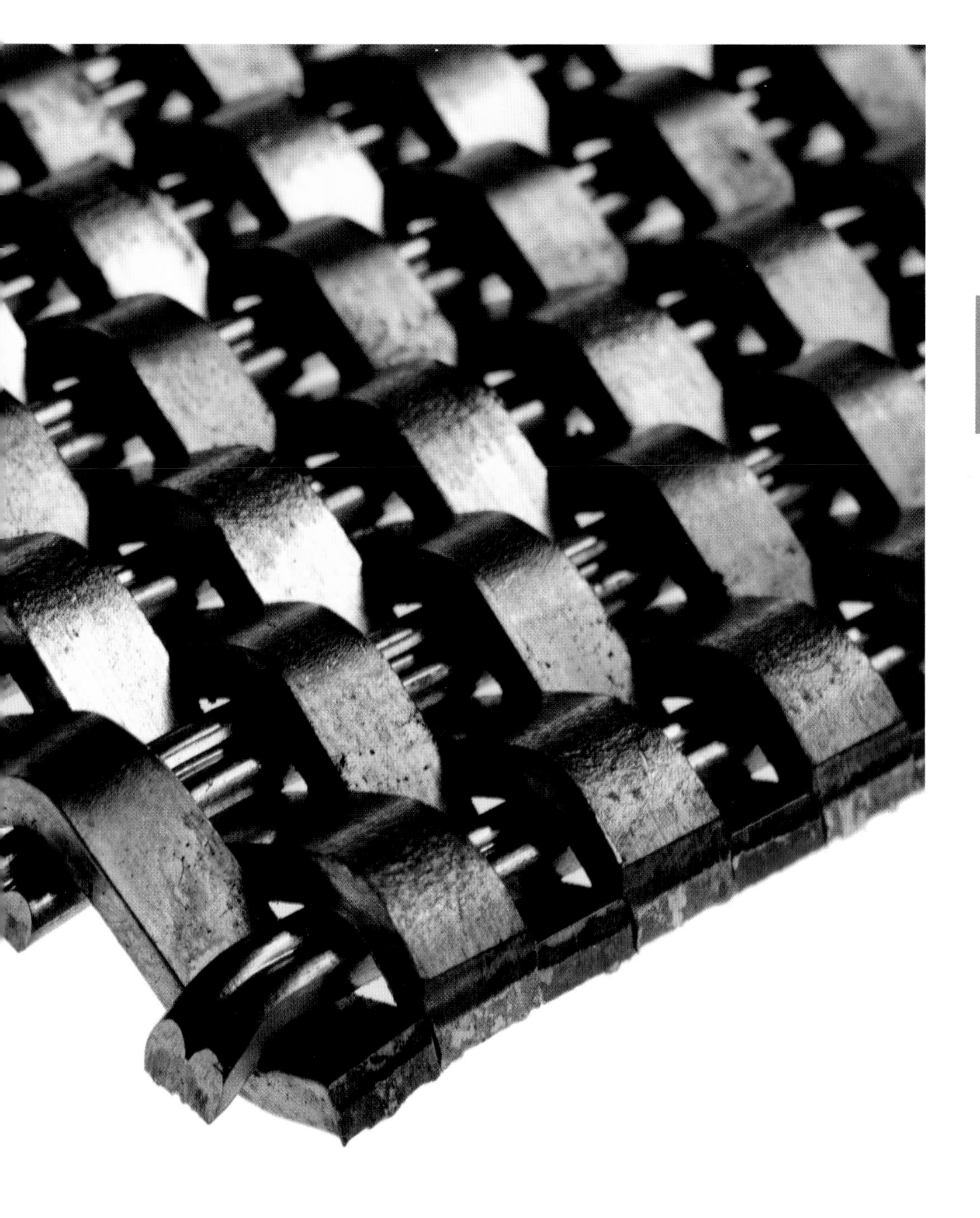

It is a two-step process. The first involves lengths of wire being crimped at specific points. This simple process is based on the wire being fed between two rollers, with teeth biting a kink into the wire at specific distances. In the second step, the long strands of crimped wire are gathered and fed into an industrial heavy-duty loom, where they are cross-layered with another set of pre-crimped wires and woven into sheets.

1 The lengths of wire are fed into the crimping machine.

2 These toothed cogs show the simple way in which crimping is achieved.

3 Weaving commences on a giant weaving machine.

4 The lengths of woven architectural mesh begin to take shape.

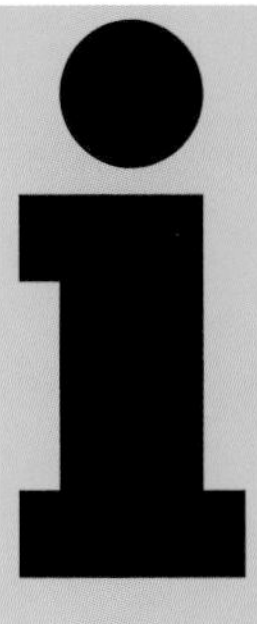

Volumes of production
From a minimum of one square metre, which may be expensive, to an unlimited number of sheets.

Unit price vs capital investment
Because of the simple wheels used in the crimping, the process does not usually require tooling. The crimping wheels themselves can be cost-effective compared with other types of industrial tooling.

Speed
Varies, depending on the type of weave.

Surface
Good finish, which can also be electro-polished (a process that removes microscopic amounts of material from the metal).

Types/complexity of shape
Flat-sheet post forming can result in infinite possibilities.

Scale
The maximum width is 2 metres. The length is restricted by the size of the manufacturer's site.

Tolerances
Not applicable.

Relevant materials
Typically uses stainless steel 316L, galvanised steel or any weavable alloy.

Typical products
Balustrades, external facades, staircase cladding, sunscreens and ceilings that allow for lighting and sprinkler systems to be fitted above.

Similar methods
Perforating expanded metal (which uses a single sheet of metal that is then pulled open to create a series of slots, and which is typically seen on the central reservation on motorways) and cable mesh – chain-link fencing that uses wire formed into a spiral, used typically in industrial security fencing.

Sustainability issues
Although entirely automated, pre-crimp weaving makes economical use of energy by crimping the wire in a separate stage prior to weaving, to ensure that the wires can be woven evenly and consistently when fed into the machine. Additionally, no heat is required during working, which significantly reduces energy use, while the woven form increases the strength and rigidity of the metal to ensure a prolonged product lifespan.

Further information
www.wiremesh.co.uk

\+

- **Adaptable, flexible production quantities.**
- **Produces a self-supporting rigid screen that can be formed and hold its shape.**

–

- **Can form only fixed-length panels, as opposed to rolls.**

Veneer Cutting
including rotary cutting and slicing

Product	Leonardo lampshade
Designer	Antoni Arola
Materials	treated wood
Manufacturer	Santa & Cole
Country	Spain
Date	2003

This simple, looped lampshade uses veneers in an unusual, decorative way that draws attention to the surprising translucency of the wood.

It is too obvious to say that trees are one of the richest sources of materials, food and shelter, but, for me, the production of veneers demonstrates the ingenuity and resourcefulness of humans in converting an object into a variety of usable forms. Peeling a tree in continuous strips to create veneers has to be one of the most economical uses for a tree and it unravels the life of the tree in the process, clearly displaying the evidence of its nutrition and lifespan.

There are two main methods for forming veneers: slicing (which involves slicing the tree – or more likely the log – along its length) and rotary cutting (which involves peeling the log in a continuous strip right into its centre until nothing is left). Rotary cutting is by far the most common form. Harvested logs are sorted according to quality to be used for veneers, pulp or conversion to plywood. Depending on the region in which the logs are gathered, they may need to be scanned for metal content. This can often be the result of bullets lodged in the trees during conflicts.

Once the logs reach the sawmill, they are cut down to the required lengths. These depend on the regional standards and whether a log will be used for veneers, or stuck together to make plywood sheets. The logs are then softened by being soaked in hot water for an average of 24 hours. This loosens the bark and relaxes the fibres in the grain, which makes the peeling process easier.

Once the bark is removed, the logs can be slowly dried before, in rotary cutting, they are set into a machine that rotates them while a cutter is introduced to slowly produce a continuous length of veneer. This length, and those produced by slicing, can be guillotined into shorter lengths.

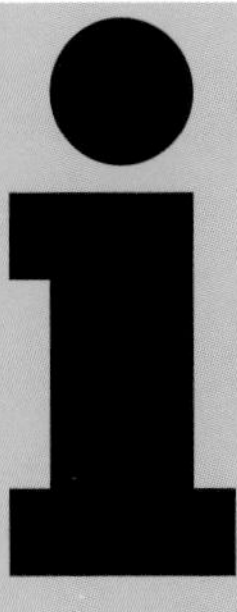

Volumes of production
Not applicable. Since this is a 'commodity', it is just produced all the time.

Unit price vs capital investment
Not applicable. Again, veneers are produced all the time, so you only pay for tools and machinery indirectly.

Speed
Once loaded into the cutter, a typical birch log (with a 300-millimetre diameter) can be completely 'peeled' in a continuous sheet in less than two minutes.

Surface
Considering that the process involves a piece of wood that in essence has been cut with a knife, the surface is fairly smooth. Finer finishing can, obviously, be achieved by sanding.

Types/complexity of shape
Thin sheet material.

Scale
The blade on the cutter can be set to cut a varying thickness of veneer from approximately 1 to 2 millimetres. The size of the sheet is determined by the width of the log and at which point the veneer is cut into smaller sheets. A typical log 300 millimetres in diameter will produce up to 15 metres of veneer.

Tolerances
Not applicable.

Relevant materials
Most tree species.

Typical products
The obvious use of veneers is in the production of various forms of plywood or veneers for laminating to board for furniture makers. However, there are also companies that laminate veneers with an adhesive and sell them as wall coverings.

Similar methods
This is a unique method of processing wood. The veneers, however, can be used to make plywood, which can be formed in a number of ways, including by bending (p.80).

Sustainability issues
Rotary cutting makes the most effective use of wood by continuously trimming the log in a full circle using all the timber, whereas the slicing process requires the log to be cut into a rectangular piece of timber prior to slicing, which creates waste. However, wood is a natural and renewable resource so constant regrowth will provide a consistent supply and prevent depletion.

Further information
www.ttf.co.uk
www.hpva.org
www.nordictimber.org
www.veneerselector.com

- **Economical use of the material.**
- **Although this is an industrial production method, it has a degree of flexibility, allowing control of the thickness of the veneer and the length and width of the final sheets.**

- **Limited to producing sheets or strips.**

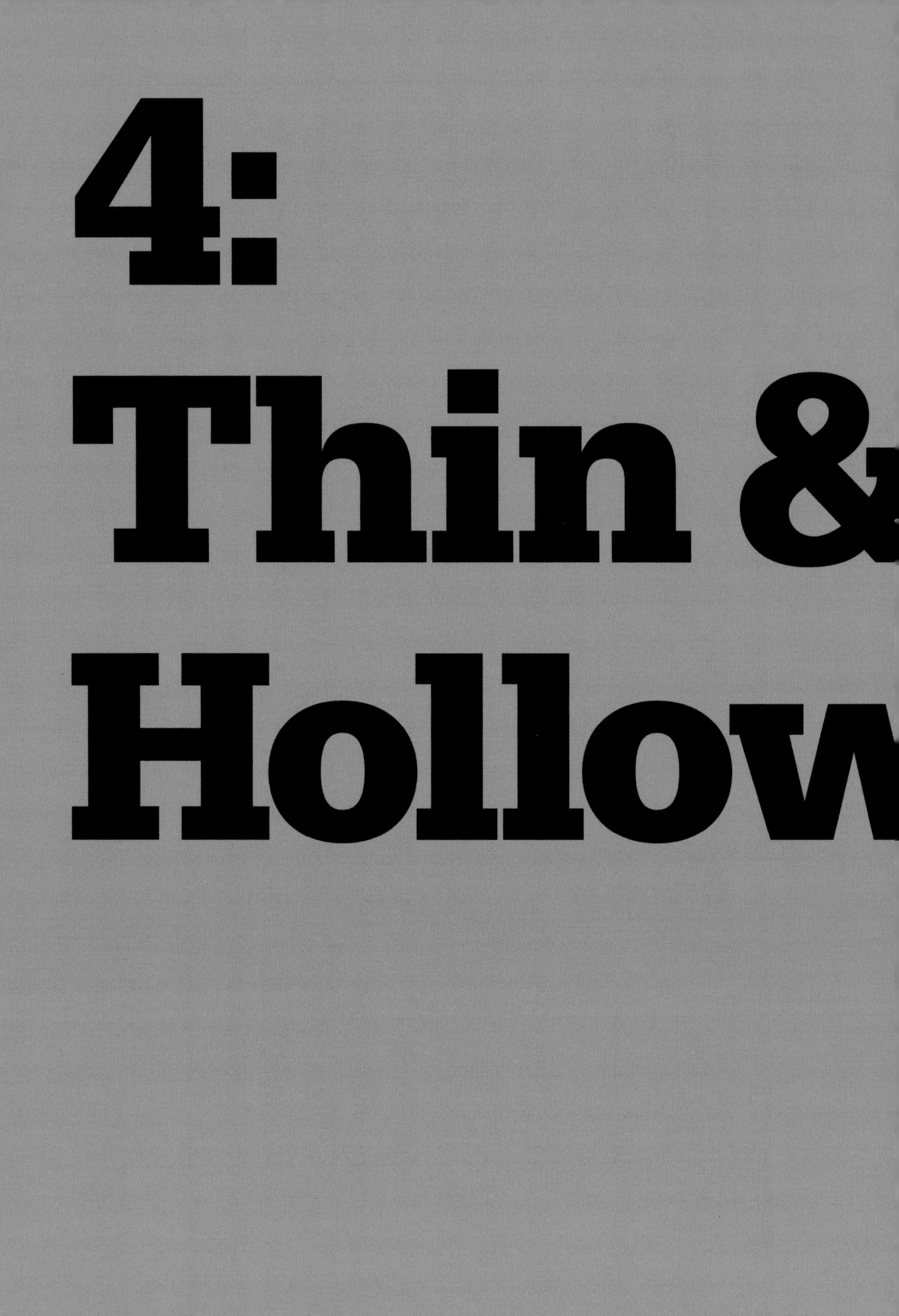
4:
Thin &
Hollow

Hollow components with a thin wall section

The longest chapter in the book, this embraces all sorts of processes for forming hollow and, generally, thin-walled shapes. It discusses the many variations of blow moulding, a process that has been used for several thousand years to produce priceless handblown glassware. The blow-moulding principle has been successfully employed in industrial mass-production, especially by the plastics industry, which spews out millions of disposable bottles for the soft drinks industry. Other forms of casting and moulding are included, from the very common rotational moulding, a form of which is used to produce chocolate Easter eggs, to the less common centrifugal casting that hurls metal or glass around a rotating drum forcing the material to attach itself to the walls, to make anything from small jewellery to huge industrial pipes.

Glass Blowing by Hand

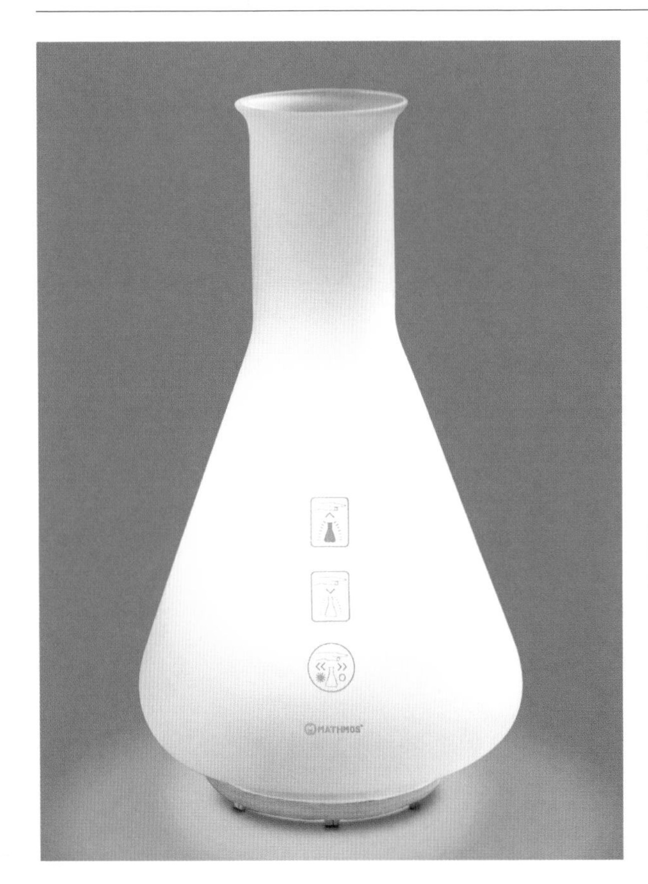

Product	Air Switch flask lamp
Designer	Mathmos Design Studio
Materials	acid-etched glass
Date	2004

Although this light was handblown, the straight sides and symmetrical shape were achieved by blowing into a mould. Usually, the shape of a handblown piece is controlled only by a series of hand tools (as illustrated in the photographs, right).

For at least two thousand years, this technique has been used to make anything from tableware to craft pieces. It involves blowing air through a metal tube to inflate a ball of gathered glass at the end of the tube. Before glass blowing, glass objects were produced by dipping a sand core in molten glass prior to rolling it against a flat surface to control the shape. Once cooled, the sand could be removed, leaving a hollow container. With the introduction of the blowing technique came a whole new set of

1 A mass of molten glass is gathered onto the end of a steel tube, ready to be blown.

2 Various hand tools are used to shape the hot glass, in this case a stack of wet fabric.

possibilities, not only in terms of shape but also in terms of widening the availability of this material.

Today, hand-blowing is still used industrially to produce a whole range of products that are blown into moulds, from lighting to wine glasses. Hand-blown glass constitutes a valuable bridge between mass-produced glassware, which requires expensive tooling and very high volumes, and individual one-off pieces.

Volumes of production
One-offs and batch production.

Unit price vs capital investment
The biggest cost is the glass-blower's labour. Assuming you want to produce a batch of identical shapes, moulds can be used. Depending on exact quantities, these will be made from materials offering varying degrees of longevity, including wood, plaster or graphite.

Speed
Completely dependent on the scale and complexity of the piece, and whether or not the glass is being blown into a mould.

Surface
Excellent.

Types/complexity of shape
For free-blown glass, virtually any shape is possible.

Scale
As big as the lungs of the glass-blower will allow, bearing in mind that the blower also needs to wrestle with the weight of the glass at the end of the tube.

Tolerances
Difficult to be precise, because it is a handmade process.

Relevant materials
Any type of glass.

Typical products
Anything from tableware to sculptures.

Similar methods
Lampworking (p.118) and machine-blown glass made using blow and blow (p.120) or press and blow (p.124) moulding.

Sustainability issues
As with all glass working, energy consumption is high due to the intense heat required over long stretches of time. But because products are shaped by hand no additional machinery is necessary, which helps to balance energy use. Any faulty mouldings or broken glass can be melted down and recycled back into the process on site to reduce material consumption.

Further information
www.nazeing-glass.com
www.kostaboda.se
www.glassblowers.org/
www.handmade-glass.com

- **Flexible enough to produce different shapes.**
- **Can be used for one-off, batch or medium-volume production.**

–

- **Units can be expensive due to labour costs.**

Lampworking Glass Tube

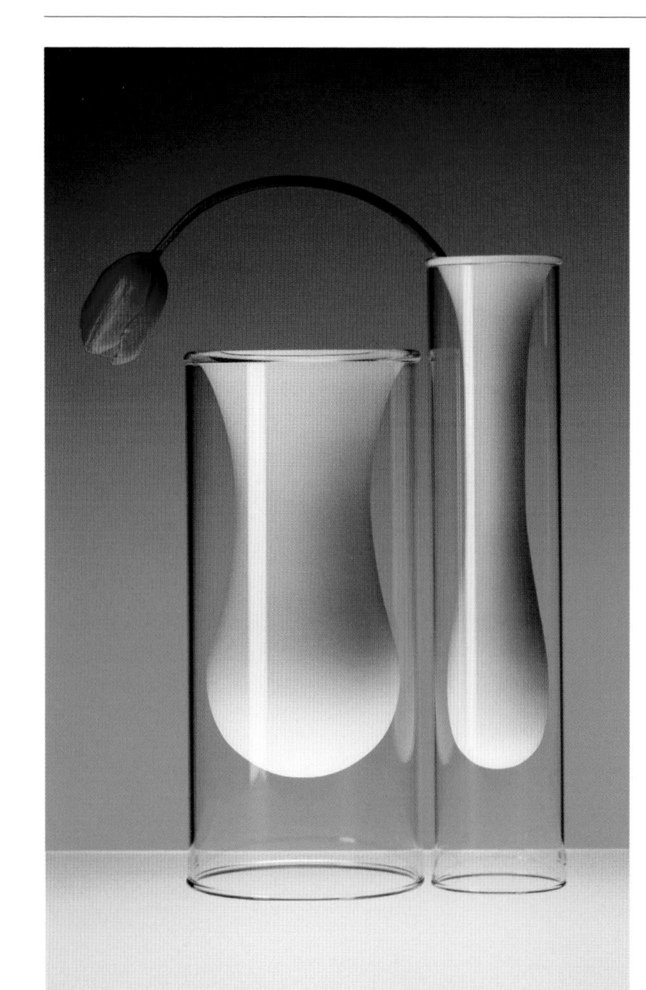

Product	thin-walled vases
Designer	Olgoj Chorchoj
Materials	borosilicate glass
Country	Czech Republic
Date	2001

These elegant vases illustrate the complexity of components that can be formed using this method. The internal opaque white form and the transparent external tube were made separately and joined together later on a lathe.

There are hundreds of ways of hand-working glass, employing both hot and cold (cutting, for example) processes for making objects without the need for tooling. Lampworking involves the localised heating of a piece of glass to allow it to be pushed, pulled and generally shaped by a skilled craftsman. The process can be seen as providing a third alternative for shaping glass somewhere between expensive hand forming and mass-production that requires tooling. It is a process that is ideally suited to short production runs.

The process starts with a hollow tube of glass, which is set into a slowly rotating lathe. Heat from a blowlamp is applied to specific areas, which are then pushed with a wooden former. Lampworking involves soft, malleable glass being pushed into shape. Depending on whether closed or open forms are required, ends can be left open or rolled round and sealed off.

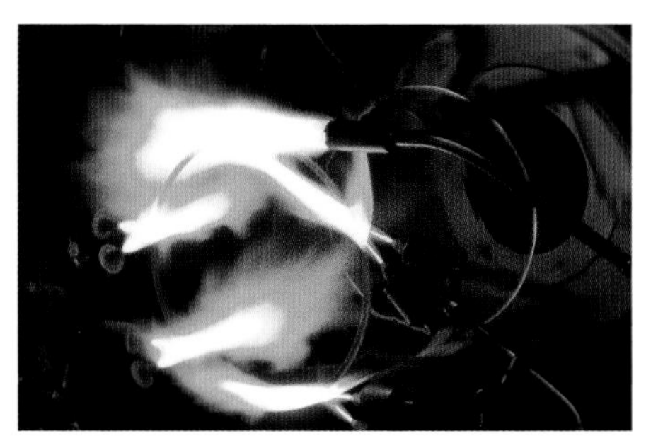

A tube of glass being locally heated while rotating on the lathe, before the wooden former is introduced.

Volumes of production
One of the best things about this type of semi-manual process is that there is no limit to the number of units that can be produced – it can be used for anything from one-offs to runs of several thousand. If you want to produce more than 1,000 units, then it might be worth considering having the product blown using a semi-automated set-up.

Unit price vs capital investment
Unit price is relatively low for a product that can be tailored and easily adapted. Capital investment is non-existent because there are no tools.

Speed
Varies, depending on the complexity of the shape.

Surface
Excellent.

Types/complexity of shape
Limits to the shape are based on symmetry because of the fact that the glass tube rotates around a single axis. However, post working of the glass once it is taken off the lathe can allow for design details to be added. Laboratory glassware is made using this method, which may give you an idea of its complexity. Wall sections are generally thin.

Scale
The scale of products is limited by the type of lathe and the skill of the craftsman.

Tolerances
Because this is a hand-worked process, tolerances are not very high.

Relevant materials
Mainly restricted to borosilicate glass.

Typical products
Anything from special laboratory apparatus and packaging, to oil and vinegar containers (the kind you find in expensive delis, where the vinegar bottle is trapped inside the oil bottle), thermometers and lighting.

Similar methods
Glass blowing by hand (p.116).

Sustainability issues
Although glass is a natural and renewable material, its production and manufacture is less eco-friendly because of the extreme heat required. However, during lampworking the glass is shaped by hand which, although time consuming, requires no machinery and therefore helps to balance this high use of energy. Additionally any glass wasted through breakage or error can be recycled back into the process to reduce material consumption and save raw materials.

Further information
www.asgs-glass.org
www.bssg.co.uk

+

- **Highly versatile process.**
- **Shapes can be varied even within the same batch.**
- **Cost-effective for experiments and prototypes.**
- **Complex shapes can be formed.**
- **This type of process for making glassware is generally used to make batch-produced products without requiring any investment in tooling.**

–

- **Not cost-effective for large production runs.**

Glass Blow and Blow Moulding

There are a number of different ways in which blowing air into, or out of, a material can be used to manufacture products, many of which are described in this book. Although varieties of blow moulding can be used for plastic (see, for example, injection blow moulding, p.129) and even – on a limited scale – metal (see inflating metal, p.76, and superforming aluminium, p.70), it remains one of the major industrial mass-production methods for making blown glass objects. The industrial blow moulding of glass today consists of two main methods: blow and blow, and press and blow (see p.124). The blow and blow method discussed here is used to make bottles with narrow necks, such as wine bottles. The term 'blown glass'

Product	**Kikkoman bottle**
Designer	Kenji Ekuan
Materials	soda-lime glass
Manufacturer	Kikkoman Corporation
Country	Japan
Date	1961

The proportions and narrow neck of this classic soy sauce bottle are typical of the blow and blow process for glass forming. The parting lines, which are just visible, show the point where the two halves of the mould have separated. The red plastic cap is injection moulded.

can, of course, also be applied to one-off handmade pieces (see glass blowing by hand, p.116), but we are talking here about the sort of large-scale process that is capable of producing hundreds of thousands of units per day.

To form a product using blow and blow moulding, a mixture of sand, sodium carbonate and calcium carbonate is carried to the top level of the factory, where it is heated to 1,550°C in a furnace that can be as large as a small living room. The molten glass is released in a series of fat sausage shapes, known as 'gobs', which are drawn down by gravity into

Volumes of production
Range from several thousand to hundreds of thousands per 24-hour period. The minimum production run to achieve an economical price is approximately 50,000 units. The weight of the glass, however, is one of the main determinants of speed, and rates of 170,000 units per day are not uncommon.

Unit price vs capital investment
This is a process for high-volume mass-production. Tooling costs are high, and production runs for glass need to last for days, on a continuous 24-hour cycle, for the products to be cost-effective.

Speed
Depending on the bottle size, machines can be set up to hold several moulds at the same time on a single machine. This can result in very high production rates, with some approaching 15,000 pieces an hour.

Surface
Excellent finish – look at any wine bottle.

Types/complexity of shape
Restricted to fairly simple forms. In large-scale glass production, the forms need to be carefully designed to allow for the easy opening of moulds – for instance, they cannot have sharp corners, undercuts or large, flat areas. The blow and blow method is, in fact, very inflexible, and you should consult a manufacturer for specific designs. Do not look at expensive perfume bottles for inspiration, because that is a different game altogether.

Scale
Because of the nature of the applications for blow-moulded products (mainly for domestic glass vessels), most manufacturing is set up for a maximum of 300-millimetre-high containers.

Relevant materials
Almost any type of glass.

Typical products
Narrow-necked wine and spirit bottles, and oil, vinegar and champagne bottles.

Similar methods
While this method is suited to making narrow-necked glass containers, press and blow moulding can make open-necked glass containers (p.124). For plastics, see injection blow moulding (p.129) and extrusion blow moulding (p.132).

Sustainability issues
Although this process has an incredibly high production rate that helps to make effective use of energy, the extreme heats required throughout the various stages of production make it incredibly energy intensive. On the positive side, glass is a natural and renewable material so has a low environmental impact while it can also be widely recycled.

Further information
www.vetreriebruni.com
www.saint-gobain-emballage.fr
www.packaging-gateway.com
www.glassassociation.org.uk
www.glasspac.com
www.beatsonclark.co.uk

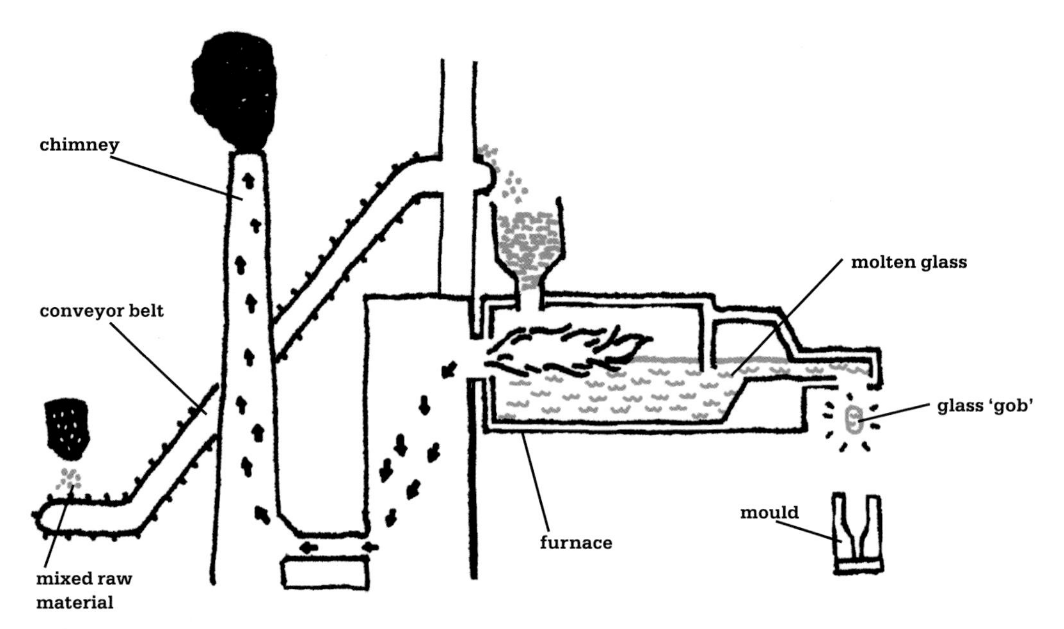

1 A mixture of sand, sodium carbonate and calcium carbonate is fed, via a conveyor belt, into a furnace at the top of the factory. Here, it is heated to make molten glass. This molten glass is released through a series of slides, and, through gravity, falls into a fat sausage shape, called a gob.

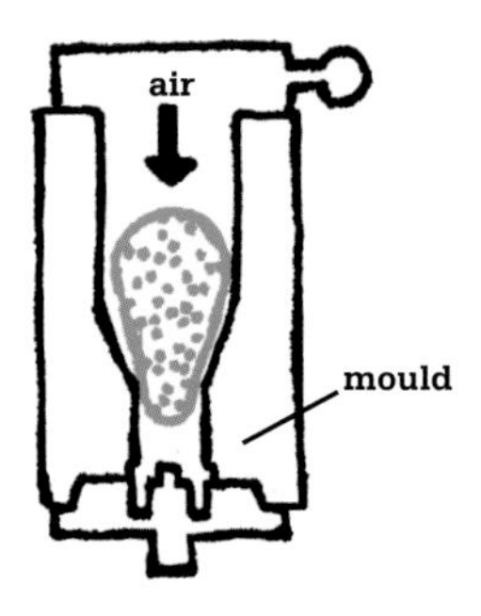

2 Blown down into the mould, the gob is the starting point for the bottle.

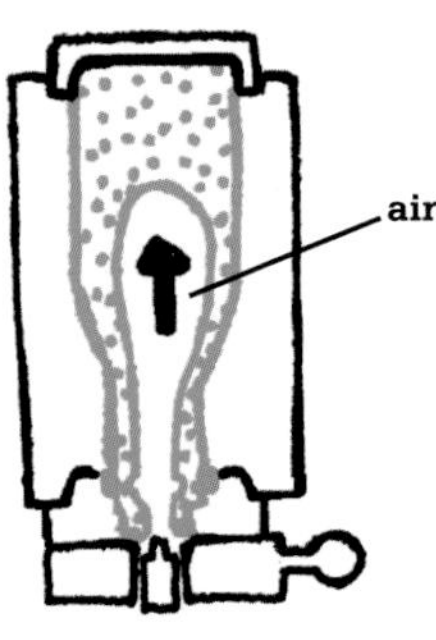

3 Air is injected into the neck to make a partially formed blank, including the neck.

4 The blank is rotated 180° and transferred to a second mould.

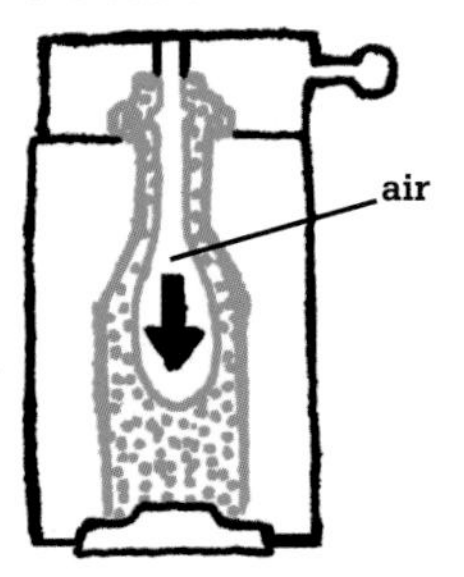

5 More air is injected.

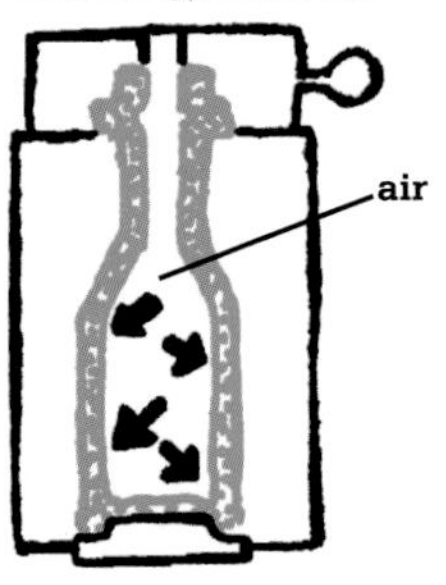

6 Air is injected until the glass is blown to form the final shape, with the glass walls at the correct thickness.

7 The glass bottle is lifted out of the mould.

the forming machines. At this stage, air is injected into the gob to partially form the bottle, including the neck. This semi-formed glass is then removed, rotated 180 degrees and clamped into a further mould. At this stage, air is injected into the mould to form the final shape. The various parts of the mould then open and the bottle is lifted onto a conveyor belt, which carries it to an annealing oven to eliminate any tension in the glass.

1 Gobs of heated glass are dropped from an elevated furnace.

2 The glass gobs are cut to length before being dropped into the mould.

3 Hot bottles leaving the mould.

4 A series of eight moulding machines feed bottles onto the production line, ready for annealing.

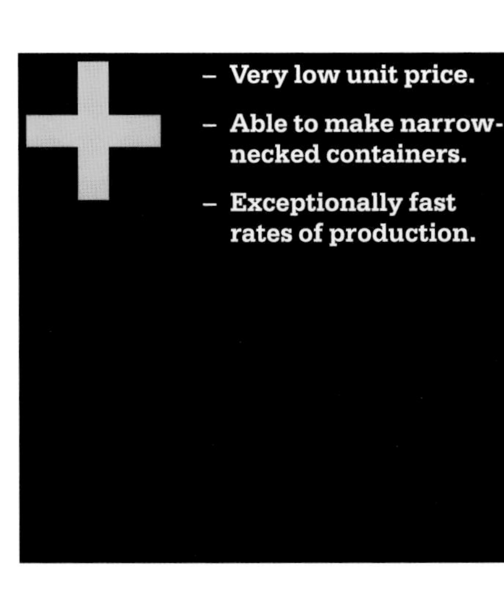

- Very low unit price.
- Able to make narrow-necked containers.
- Exceptionally fast rates of production.

- Versatility is very low in this high-volume method of production.
- Very high tooling costs.
- Demands very high volumes.
- Limited to fairly simple hollow forms.
- Adding colour to glass can be expensive as it involves ‘running through’ colours at the end of production to ensure that there is no bleeding between colours.

Glass Press and Blow Moulding

A form of industrial glass blow moulding, the technique known as 'press and blow' is used to make wide-mouthed containers such as jam jars, rather than the narrow-necked items such as wine bottles, that are made with the blow and blow process (see p.120). The main difference between the techniques occurs during the moulding process. Instead of being blown, to create wide-mouthed vessels the 'gob' of glass is pressed onto a male former inside the mould cavity. This can speed up production cycles and allows greater control in the distribution of the glass, so that a thinner wall can be achieved. After the objects have been formed, the production line pushes them into an annealing furnace where, over the period of an hour, the objects are slowly cooled to room temperature, thus eliminating any tension in the glass.

Inside the factories, machines shoot out glowing, molten gobs of glass that look like shafts of light falling into the cavities of the empty moulds. This process has none of the theatre and craftsmanship of hand-blown glass: the automated, greasy, noisy, steaming machines can turn out hundreds of thousands of bottles per day with just a handful of men watching over this vast production.

Compared with the blow and blow process, which can produce over

Product	**storage jar**
Materials	soda lime glass with thermoplastic elastomer (TPE) seal
Manufacturer	Vetrerie Bruni
Country	Italy

The open necked shape of this jar is a typical example of a product for which you would have to consider press and blow moulding in preference to blow and blow moulding (see p.120).

350,000 narrow-necked units per day, this process can churn out 400,000 units the size of, for example, jam jars. When it comes to small 'press and blow' bottles, however, the machines can pump out up to 900,000 units of, say, small pharmaceutical bottles per day, running on a continuous 24-hour cycle. Uninterrupted production runs for some food packaging can last up to ten months, just producing the same objects over and over again.

Volumes of production
Range from several thousand to hundreds of thousands per day. This level of high-volume production is usually determined by time, rather than by numbers of units produced per hour. It may take up to eight hours for production to be in full swing, so a minimum production cycle is likely to be around three days, with machines running without interruption.

Unit price vs capital investment
As with the similar process of blow and blow moulding (see p.120), this is a process only for high volume mass-production. Tooling is prohibitively expensive unless you have production runs of several tens of thousands of units.

Speed
The press and blow method is generally slightly faster than blow and blow glass production, though they have in common the fact that the weight of the glass is a determining factor for speed. Rates of 250,000 units per day for a typical large cooking-sauce jar are fairly standard.

Surface
Just look at a jam jar and you can see the excellent finish. However, just as with blow and blow bottles, the witness lines will need to be taken into account if labels are to be added.

Types/complexity of shape
Restricted to fairly simple forms with wide, open necks. In large-scale glass production these forms cannot have sharp corners, undercuts or large, flat areas, all of which would make releasing them from the mould difficult. Compared with blow and blow moulding, press and blow allows a greater degree of control over the thickness of the glass.

Scale
As with blow and blow, manufacturing is set up for a maximum of 300-millimetre-high containers.

Relevant materials
Almost any type of glass.

Typical products
Open-necked jam jars and spirit bottles, open-necked pharmaceutical and other containers, and food packaging.

Similar methods
For glass, blow and blow moulding (p.120), lampworking (p.118) and glass blowing by hand (p.116). For plastics, plastic blow moulding (p.127) and extrusion blow moulding (p.132).

Sustainability issues
Similar to blow and blow moulding, the extreme heats used throughout various stages of production amount to an exceedingly high energy consumption. Yet, the exceptionally high production rate and fast cycle times are optimised to make economical use of this energy, whilst the recycling of glass back into the process helps to reduce the use of raw materials.

Further information
www.vetreriebruni.com
www.britglass.org.uk
www.saint-gobain-conditionnement.com
www.beatsonclark.co.uk

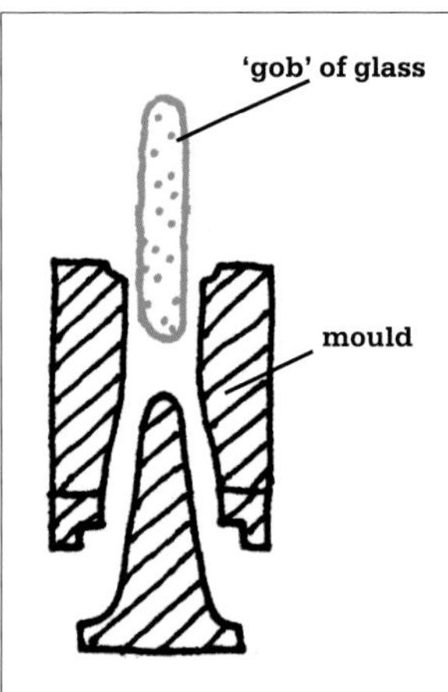

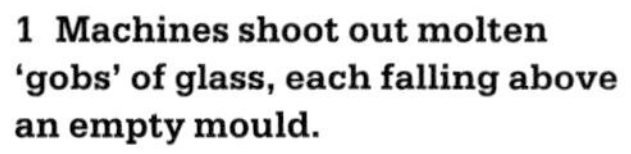

1 Machines shoot out molten 'gobs' of glass, each falling above an empty mould.

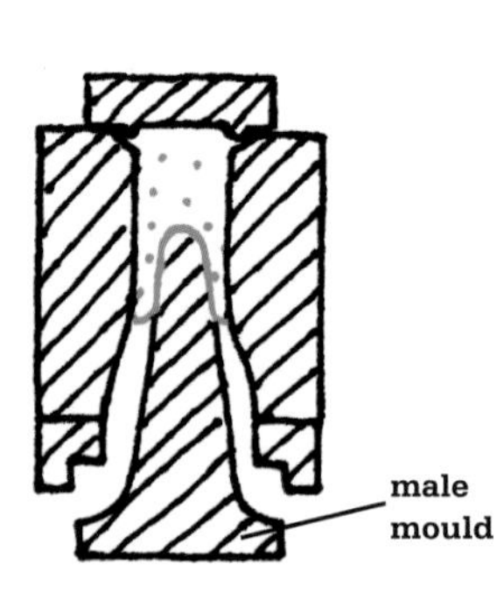

2 The male part of the mould begins to shape the glass as it falls.

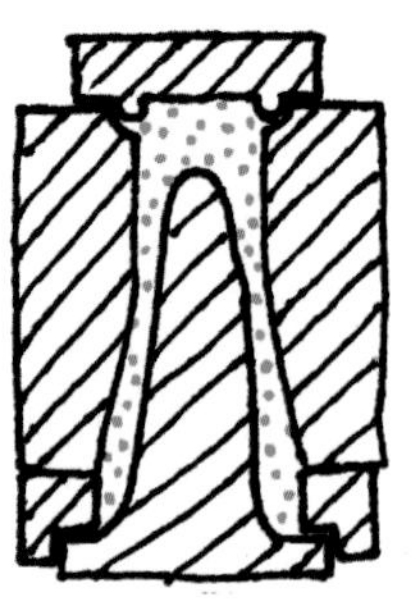

3 The soft glass is pressed right down into the mould to form a blank.

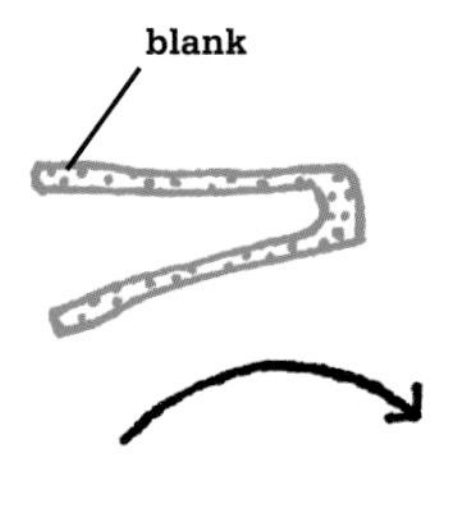

4 The blank is rotated 180 degrees.

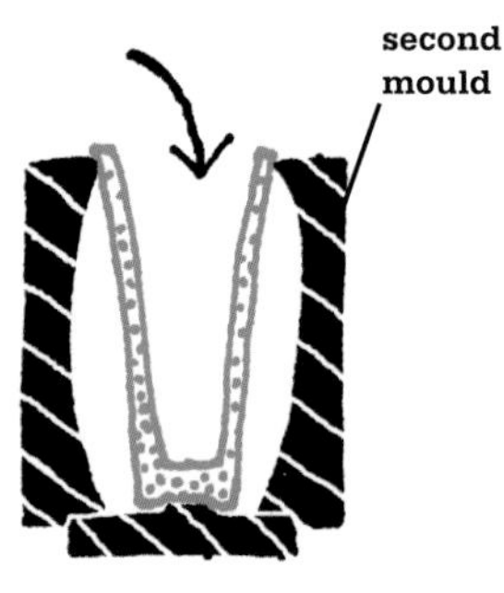

5 The blank is transferred into a second mould.

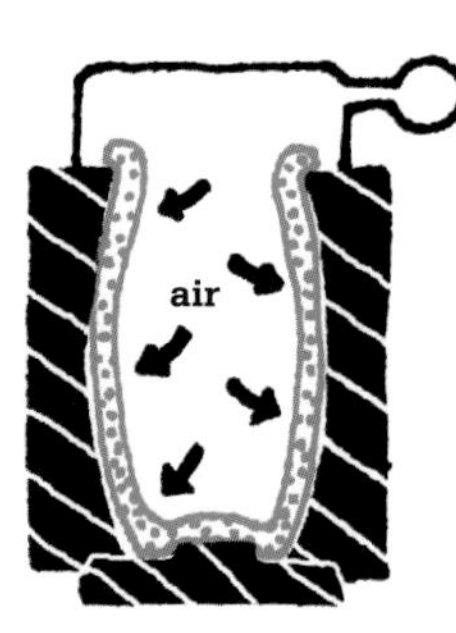

6 Air is used to blow the glass right into the mould to form the final shape.

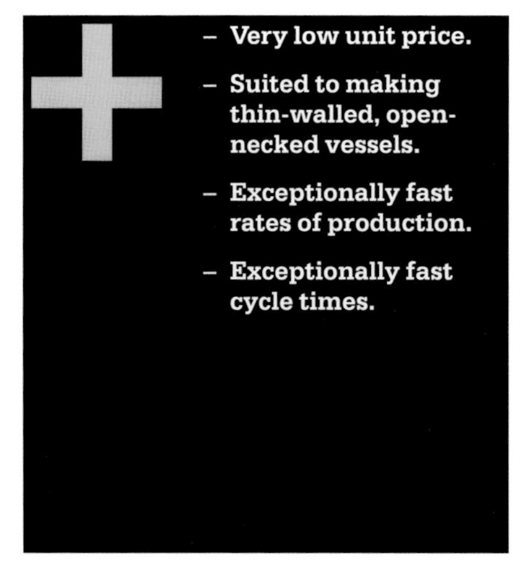

– Very low unit price.
– Suited to making thin-walled, open-necked vessels.
– Exceptionally fast rates of production.
– Exceptionally fast cycle times.

– Very high tooling costs.
– Limited to fairly simple hollow forms.
– Adding colour to glass can be expensive, as can running through colours at the end of production to clean out the machines.
– Demands high volumes in order to be economical.

Plastic Blow Moulding

Blow moulding is an umbrella term that describes one of the major industrial mass-production methods for producing a whole host of hollow products. In one sense it is unusual, because it is a process that can be used for moulding plastic containers as well as glass bottles (see glass blow and blow [p.120] and glass press and blow [p.124] moulding).

There are several forms of blow moulding suitable for plastics, including injection blow moulding and injection stretch moulding (see p.129), and extrusion and co-extrusion blow moulding (see p.132). All have differing potential to create shapes, but, in simple terms, all of them involve a process that is like blowing a balloon into a mould to form a shape. The process starts with a pre-form being fed into a two-part mould. The closing of the mould snips the material to an appropriate length, forming a seal at one end of the plastic. This pipe-like form is fed into a second mould where air is blown into it, forcing the plastic to expand against the mould cavity to form the final shape, after which the mould opens and the part is released.

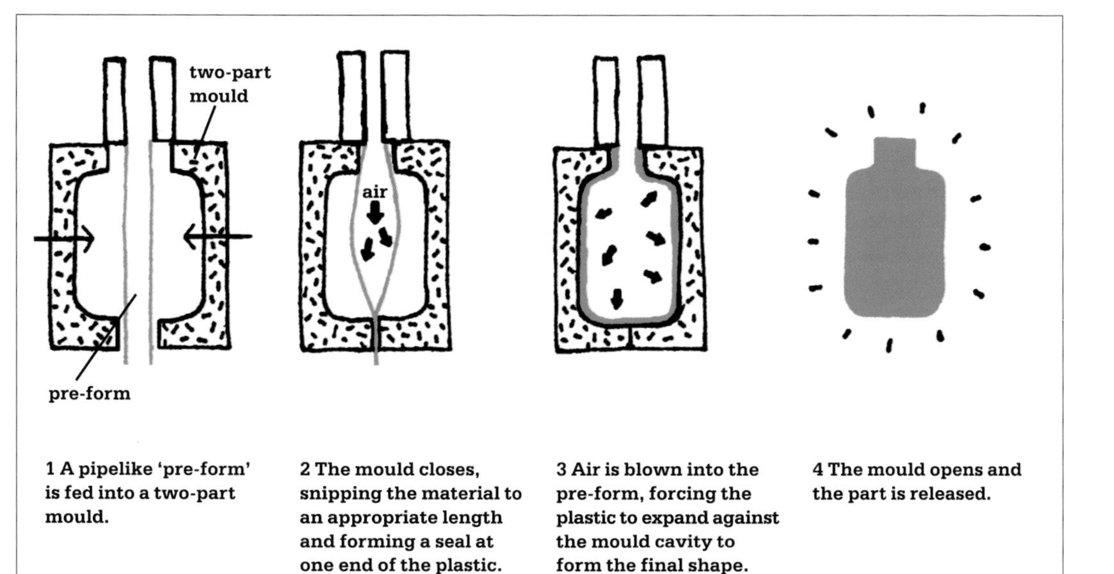

1 A pipelike 'pre-form' is fed into a two-part mould.

2 The mould closes, snipping the material to an appropriate length and forming a seal at one end of the plastic.

3 Air is blown into the pre-form, forcing the plastic to expand against the mould cavity to form the final shape.

4 The mould opens and the part is released.

Volumes of production
Depending on the size and the material, blow moulding can be an extremely rapid form of production, with an output of from approximately 500 units per hour to over a million units per 24-hour cycle. To get the most out of the process, in terms of cost savings, production should run into the hundreds of thousands.

Unit price vs capital investment
Unit prices for most standard blow-moulded parts are very low, which is best comprehended by looking at the volume of cheap products and packaging that is produced with this process. This economy of scale is counterbalanced, of course, by the extremely high tooling costs.

Speed
Small containers can be produced on multi-cavity moulds to yield approximately 60,000 small (less than, say, 700 millilitres) polyethylene terephthalate (PET) bottles per hour.

Surface
Excellent finish, but parting lines remain down the length.

Types/complexity of shape
Depending on the specific process, blow-moulded shapes are generally simple and rounded. Although products can be produced with no draft angles, manufacturers prefer a small draft.

Scale
From small cosmetics bottles to parts that weigh over 25 kilograms.

Relevant materials
The typically waxy, high-density polyethylene (HDPE) is one of the most common materials used for this process. Other materials include polypropylene, polyethylene, polyethylene terephthalate (PET) and polyvinyl chloride (PVC).

Typical products
The chances are that in the average household you will have at least one large cupboard full of different plastic containers that are blow moulded. Basically, blow-moulded items include everything from plastic milk cartons and shampoo bottles to toys, toothpaste tubes, detergent bottles, watering cans and – outside the home – car fuel tanks.

Similar methods
Stretch moulding, extrusion blow moulding (p.132), injection blow moulding and co-extrusion blow moulding (p.132).

Sustainability Issues
A highly automated process that as a result of very fast cycle times maximises the use of heat and electrical energy to produce a very precise quantity of material and optimised use of energy. PET, which is one of the main materials for blow moulding, is among the most widely recycled plastics.

Further information
www.rpc-group.com
www.bpf.co.uk

+

- **Very low unit price.**
- **Exceptionally fast rates of production.**
- **Details, such as threads, can be moulded in.**

–

- **High tooling costs.**
- **Demands high volumes in order to be cost-effective.**
- **Limited to fairly simple hollow forms.**

Injection Blow Moulding
with injection stretch moulding

Injection blow moulding is most easily described as being a subdivision of plastic blow moulding (see p.127), the process that works on the same principle as blowing up a balloon, but into a mould that forms the shape.

As the name implies, this is a two-step moulding process that offers a number of advantages over other forms of blow moulding because it is possible to create far more complex shapes around the neck of the moulded part. A hollow pre-form is made using injection moulding (see p.196), which allows for the moulding of a complex thread at the neck. The pre-form is placed into the mould cavity where it is blown with air, forcing the plastic against the mould cavity.

Using an injection-moulded pre-form means that this method offers a greater degree of stability and control over the shape than extrusion blow moulding (see p.132), although the choice of suitable materials is more limited.

Injection stretch moulding is a method used for high-end products (such as bottles) made from polyethylene terephthalate (PET) which uses a rod to stretch a pre-form into the mould before blowing.

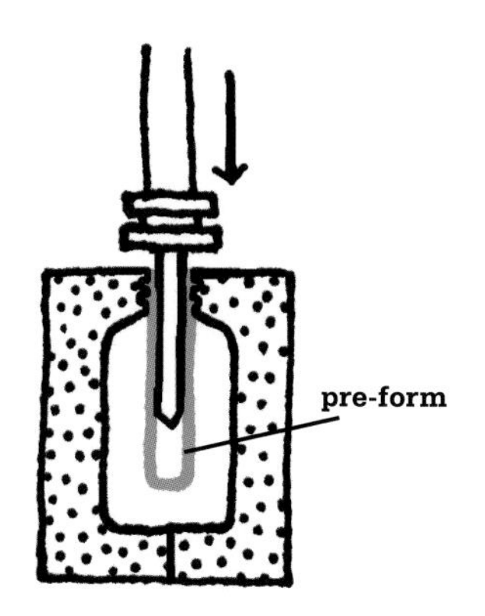

1 An injection-moulded pre-form is placed in the mould.

air

2 Compressed air is injected, blowing the pre-form into the mould cavity to form the final shape.

3 The mould opens and the part is released.

Product	injection-moulded pre-form (left) and blow-moulded bottle (right)
Materials	polyethylene terephthalate (PET)
Country	Germany

This pre-form and the resulting blown bottle show, in very simple terms, how straightforward the process is that forms the billions of plastic bottles that litter our urban landscape. The advantage of using injection moulding is demonstrated by the detailed thread that has been formed around the neck of the pre-form.

Product	Sparkling Chair
Designer	Marcel Wanders
Materials	polyethylene terephthalate (PET)
Manufacturer	Magis
Country	Italy
Date	2010

The Sparkling Chair is a great illustration of how a material and production method can be translated from packaging into a completely new area of furniture.

Volumes of production
Injection blow moulding is ideally suited to high-volume production, which often runs into millions of units.

Unit price vs capital investment
Costly tooling, for both the injecting and blowing parts of the process, as well as substantial set-up charges. However, unit prices can be extremely low because of the volumes produced, and this justifies the high initial costs.

Speed
The various forms of blow moulding are difficult to pin down in terms of speed of production due to variables such as part size and the number of mould cavities in operation. A typical 150-millilitre bottle, however, can be produced by injection blow moulding in an eight-cavity mould at the rate of 2,400 units per hour.

Surface
Excellent finish.

Types/complexity of shape
Injection blow moulding is suited to fairly simple shapes, which have a large radius and consistent wall thicknesses over the whole product.

Scale
Typically used for containers of less than 250 millilitres.

Relevant materials
Compared with extrusion blow moulding (see p.132), this method is suited to more rigid materials such as polycarbonate (PC) and polyethylene terephthalate (PET). It is, however, often used for non-rigid materials, such as polyethylene (PE).

Typical products
Small shampoo, detergent and other bottles.

Similar methods
Extrusion blow moulding (p.132) for plastic, and press and blow moulding (p.124) for glass.

Sustainability issues
As opposed to other forms of plastic blow moulding, the pre-form is heated twice: first during production of the pre-form and then when it is blown into the final product, thus doubling on energy use. Virtually no waste is created during injection blow moulding and cycle times are very fast with optimised use of material and energy. The process is often used to produce disposable PET packaging; as one of the most widely recycled plastics PET can be reprocessed to avoid landfill.

Further information
www.rpc-group.com
www.bpf.co.uk

- **Very low unit price.**
- **Exceptionally fast rates of production.**
- **Suited to small containers.**
- **Allows greater control over neck design, weight and wall thickness than other blow-moulding methods.**

- **Higher tooling costs than in extrusion blow moulding (see p.132).**
- **Demands high volumes.**
- **Limited to fairly simple hollow forms.**

Extrusion Blow Moulding

with co-extrusion blow moulding

Extrusion blow moulding is part of the plastic blow moulding group of processes (see p.127). In this particular method, the plastic is extruded (see extrusion, p.96) into a sausage shape known as a 'lug' and pinched into short lengths as it drops into the mould cavity. Here, it is blown with air, forcing the plastic against the mould cavity to form the final shape. The process leaves excess 'pinched' material (a 'tail'), which must be removed – though evidence of it can be seen in the finished product, on the underside of any shampoo bottle, for example.

In co-extrusion blow moulding different materials are combined to form a multi-layered product.

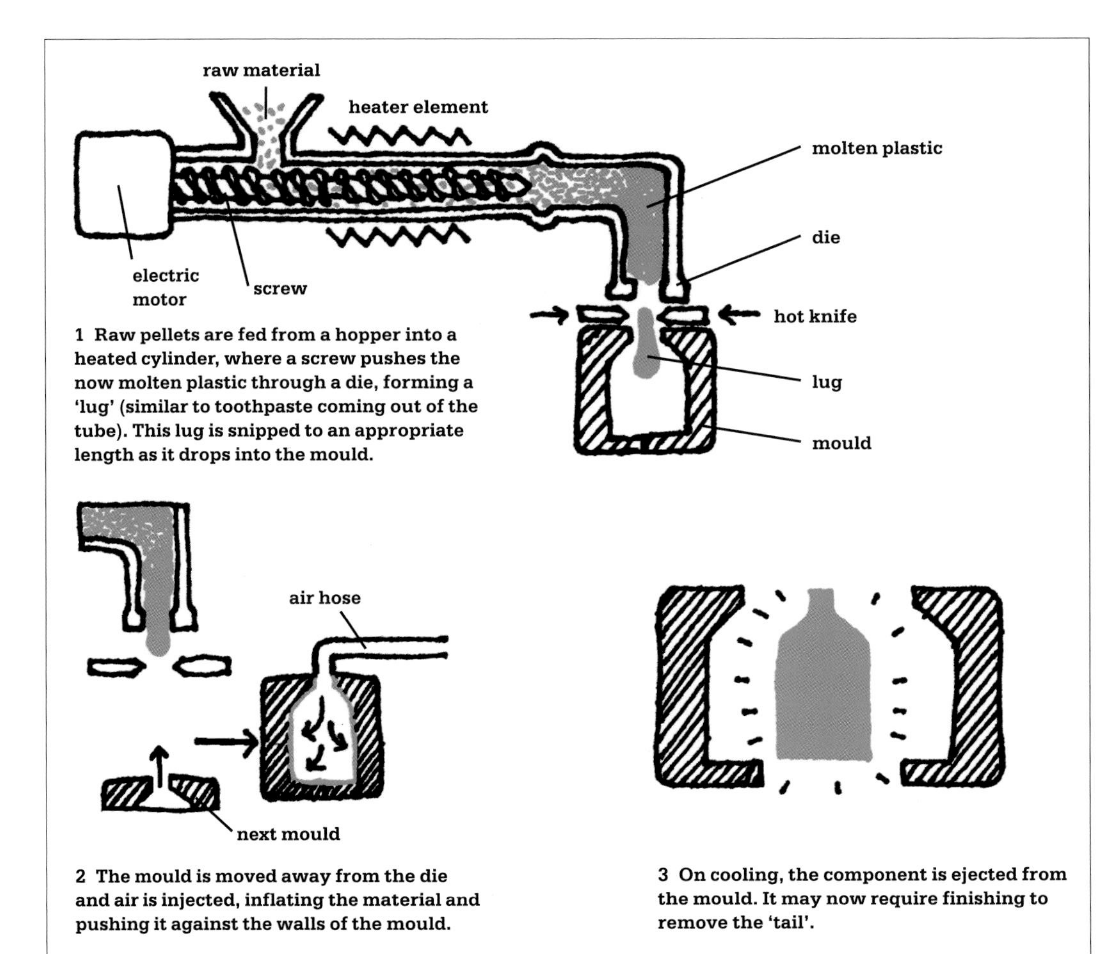

1 Raw pellets are fed from a hopper into a heated cylinder, where a screw pushes the now molten plastic through a die, forming a 'lug' (similar to toothpaste coming out of the tube). This lug is snipped to an appropriate length as it drops into the mould.

2 The mould is moved away from the die and air is injected, inflating the material and pushing it against the walls of the mould.

3 On cooling, the component is ejected from the mould. It may now require finishing to remove the 'tail'.

Volumes of production
Unlike injection blow moulding (see p.129), which offers the possibility of production runs extending into the millions, extrusion blow moulding can be used on much shorter runs, sometimes as low as 20,000.

Unit price vs capital investment
Although lower in cost than injection blow moulding (by about a third), it is still an expensive set-up.

Speed
As with other, similar methods, the production rate is determined by the weight of the part: a typical 5-litre container can be produced at a rate of 1,000 per hour (using a single machine, with four moulds running concurrently). Blow-moulded milk bottles of the sort found in supermarkets can be made at a rate of around 2,000 units per hour.

Surface
Excellent finish.

Types/complexity of shape
Extrusion blow moulding is suited to the production of larger and more complex shapes than injection blow moulding, notably the integrated handles on plastic milk containers, or on the large petrol containers you can find at filling-stations.

Scale
Although it is capable of producing products such as shampoo bottles, extrusion blow moulding is also suitable for small runs and can be used to make products at the larger end of the blow-moulding scale, typically over 500 millilitres.

Relevant materials
Polypropelene (PP), polyethylene (PE), polyethylene terephthalate (PET) and polyvinyl chloride (PVC).

Typical products
Extrusion blow moulding is best suited to larger products, which might typically include toys, oil drums and car fuel tanks, and large detergent bottles.

Similar methods
Injection blow moulding (p.129) and rotational moulding (p.137).

Sustainability issues
A small amount of excess plastic produced from runners where the molten plastic was fed in is trimmed from each mould. This waste can be heated into a molten state again and recycled back into the process to reduce material consumption and prevent waste. Additionally, at the end of the product's lifespan the plastic can be recycled to reduce the use of raw and non-renewable materials.

Further information
www.rpc-group.com
www.bpf.co.uk
www.weltonhurst.co.uk

+

- **Very low unit price.**
- **Fast rates of production.**
- **Suited to large containers of over 500 millilitres.**
- **Compared with injection blow moulding (see p.129), extrusion blow moulding is capable of producing more complex shapes.**
- **Lower tooling costs than for injection blow moulding.**

–

- **Demands high volumes.**

Dip Moulding

Dipping a shape into a material that has been melted (or is in an otherwise liquid state) is possibly one of the oldest methods of forming shapes. It is also one of the simplest techniques to understand, and, in terms of tools and moulds, it is one of the cheapest methods of producing plastic products.

To be presented with a ceramic former, such as the one illustrated here, is to be given a gem from the usually hidden world of manufacturing. Artists (particularly Rachel Whiteread in her award-winning 1993 concrete sculpture *House*) have often explored the negative spaces within our environments. In a similar way, these little gems give us a unique view into the world of production from an angle that is rarely seen. The bulbous shape triggers a small flash of recognition,

Product	**balloon former (far left) and balloon (left)**
Designer	Michael Faraday created the first rubber balloon in 1824
Materials	earthenware ceramic former; latex balloon
Manufacturer	Wade Ceramics Limited (balloon former)

The simple ceramic former perfectly illustrates the principle behind dip moulding, showing how hollow products – such as this party balloon – are made.

but you cannot quite put your finger on what it is until someone tells you that the shapes are ceramic formers for making balloons.

In principle, the process of dip moulding is incredibly straightforward. As the name suggests, you simply dip a former into a liquid polymer bath, let it cure and peel it off. In reality, it is a little bit more complex than that, because dip moulding is a process that can be adapted to many different materials and set-ups, although the basic idea stays the same.

An automated production line showing the dipping of ceramic formers to make rubber gloves.

A vat of sky-blue latex being used to produce party balloons.

- Highly cost-effective for short production runs.
- A prototype former and sample mouldings can be produced in a matter of days.

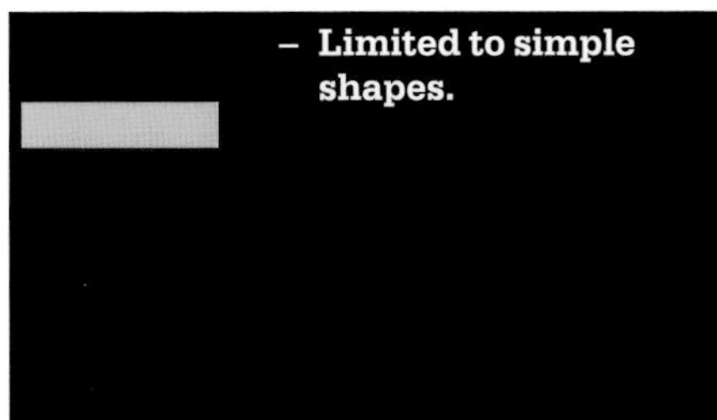

- Limited to simple shapes.

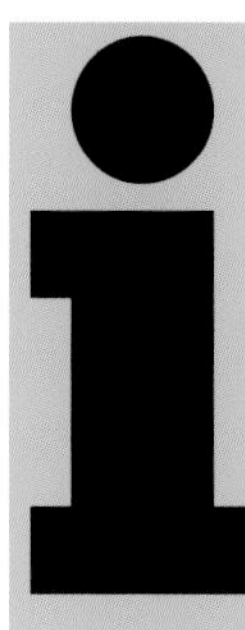

Volumes of production
From batch production to high-volume mass-production.

Unit price vs capital investment
One of the least expensive ways of mass-producing plastic components, with reasonably cheap tooling and easy-to-produce samples, while still allowing for cost-effective unit parts.

Speed
This process involves many steps, including pre-heating of the former, dipping, curing and finally peeling the finished moulding from the former, which makes for a slow process if performed manually. Complex mouldings may take up to 45 minutes to complete, while the production of very simple shapes, such as end caps (for example, simple bicycle-handlebar grips) can be fully automated and may take only 30 seconds.

Surface
The exterior of the component is determined by the natural state of the material, and may have a small nipple as evidence of the dripping polymer from the mould.

Types/complexity of shape
Soft, rubbery, flexible, though simple, forms. Products must be shaped in such a way that they can be unpeeled from the mould.

Scale
The scale of dip mouldings is theoretically only limited by the size of the bath containing the polymer, but generally mouldings range from 1-millimetre diameter end caps to 600-millimetre industrial pipe covers.

Tolerances
Dip moulding does not achieve a high level of accuracy, apart from on the internal dimensions.

Relevant materials
Because of the nature of the process, which involves the former being 'undressed' as the part is removed, it is limited to soft materials and parts that can be stretched over the moulds, including PVC, latex, polyurethanes, elastomers and silicones.

Typical products
A whole range of flexible and semi-rigid products, from kitchen and surgical gloves to balloons and those soft, waxy plastic handlebar grips for children's bikes.

Similar methods
An economical alternative to plastic blow moulding (p.127) and rotational moulding (p.137).

Sustainability issues
Heat is required to keep the polymer bath in its molten state throughout processing so dip moulding is energy intensive. Furthermore, some plastics such as latex and silicone are often not widely recyclable. On a brighter note, latex products such as balloons can, in fact, be composted, which prevents the material entering the waste stream.

Further information
www.wjc.co.uk
www.uptechnology.com
www.wade.co.uk
www.qualatex.com

Rotational Moulding
AKA Roto Moulding and Rotational Casting

Rotational moulding is all about making things that are hollow. If you have ever wanted to know how chocolate Easter eggs are made, then the answer lies in this method of production. One of the interesting things about rotational moulding is that the soft and rounded products that are typical of this method very much take their aesthetic from the limitations of the process. This is quite unlike injection moulding (see p.196), which uses pressure to inject material into the mould, producing sharp edges and fine detail. Roto moulding, as it is sometimes known, uses only heat and the rotation of a mould to form parts and thus lacks the fineness of pressure-formed parts.

In a sense, rotational moulding is based on a similar idea to ceramic slip casting (see p.140). In both methods, a liquid material is built up on the internal cavity of a mould, allowing the manufacture of hollow parts. It is a simple, four-stage process, which begins with adding powdered polymer to a cold die. The amount of powder in relation to the size of the die determines the wall thickness of the final component. The second stage involves the die being uniformly heated inside an oven, while simultaneously

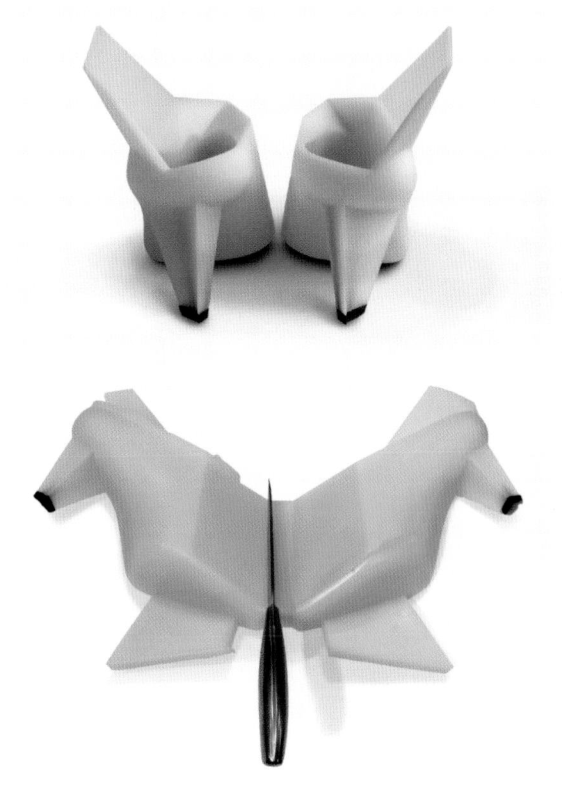

Product	Rotationalmouldedshoe
Designer	Marloes ten Bhömer
Materials	Polyurethane rubber and stainless steel
Manufacturer	Marloes ten Bhömer
Country	UK
Date	2009

The shoes demonstrate a production process being transferred into a completely new type of product. The image above shows how the two parts are separated (not actually with a knife) and rejoined to make the shoe.

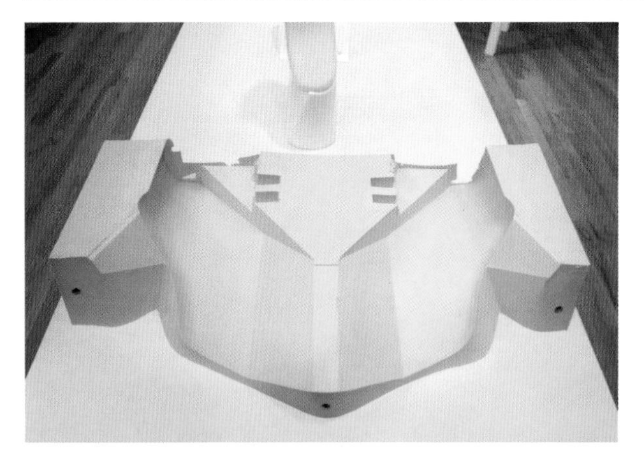

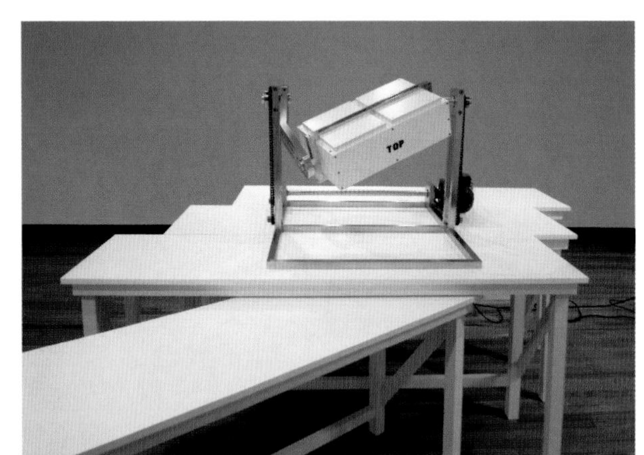

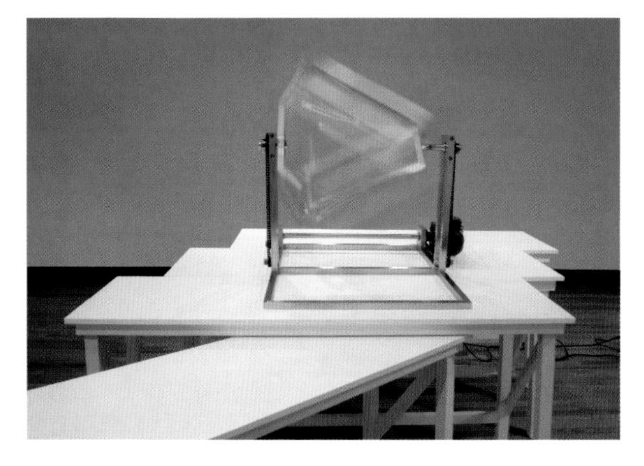

These images show (top) close-up of half of the rotational moulding tool; (middle and above) table-top rotational moulding.

being slowly rotated around two axes. This allows the polymer (which is now liquid) to tumble around the inside of the die, where it builds up on the walls and creates a hollow form. Finally, while the die is still rotating, it is cooled using air or water before the component is removed.

- Ideal for hollow shapes.
- Suitable for low-volume production.
- Simple process.
- Allows for cost-effective production of large components.

- Not suitable for making small, precise components.

Volumes of production
From batch production to high-volume mass-production.

Unit price vs capital investment
Less expensive to set up and operate than injection moulding (see p.196). Because there is no pressure involved, moulds are simpler and cheaper. Unit costs are still very low.

Speed
This is affected by the size of the component and the wall thickness, both of which affect the cooling-cycle time. Some components, such as plastic drums for storing liquids, may require entry and exit holes for the liquid to be cut by hand.

Surface
The internal surface may reveal the swirls of the plastic as it was being formed, similar to the swirls of chocolate you can see on the inside of an Easter egg. The surface that is in contact with the mould is of much higher quality. While it may not be possible to achieve a super-glossy finish, matt finishes can be built into the mould to hide small defects. Inserts with graphics on can also be moulded into parts.

Types/complexity of shape
Adaptable to a range of shapes. Even undercuts are possible. Wall thickness should be kept uniform, between approximately 2 and 15 millimetres. Unlike with other processes, there can be a build-up of material in corners which makes them the strongest part of the component.

Scale
Starting with chocolate eggs, it is possible to manufacture hollow products up to 7 metres long by 4 metres wide, such as panels for workmen's temporary huts.

Tolerances
Compared with other plastic moulding methods, tolerances are low due to shrinkage, cooling rates and the wall thickness, which varies slightly across the moulding.

Relevant materials
Polyethylene, which has that Edam-cheesy feel, is a common material for rotational moulding. Other resins can also be used, including acrylonitrile butadiene styrene (ABS), polycarbonate, nylon, polypropylene and polystyrene. Reinforcement fibres can also be introduced to increase strength in the final component.

Typical products
Chocolate eggs, plastic road-traffic cones, portable toilets, tool cases, large toys that take up half your living room, as well as many other hollow products.

Similar methods
Centrifugal casting (p.161) is a similar process for plastics, but it is not widely available and can only produce small parts. Also blow moulding in all its forms (pp.120–33), and dip moulding (p.134).

Sustainability issues
As with most plastic processing, high temperatures are needed to melt the plastic, which makes the process quite energy intensive. However, this is a pressureless process. It is difficult to control the precise wall thickness and therefore the amount of material that is needed. Any faulty mouldings produced can be melted down and reused in the process again to ensure that material consumption is kept at its lowest.

Further information
www.bpf.co.uk
www.rotomolding.org

Slip Casting

This is a manufacturing process that is just as likely to be used in a college foundation art and design course as in the industrial workshops of Wedgwood® or Royal Doulton. In slip casting, ceramic particles are first suspended in water to form 'slip', which is something like the colour and consistency of melted chocolate. This slip is tipped into a plaster mould. Because the dry plaster mould is porous, the liquid is absorbed from the outer layers of the slip, leaving a coating of leathery and hard ceramic

Product	**Teapot, before finishing**
Materials	bone china

It is often the unfinished article that best reveals the production process, rather than the finished product. This image was taken while the clay was still wet, before the excess material at the top is trimmed off. The parting lines, where the two halves of the mould met, are still visible on the sides of the teapot.

on the inner surface of the mould. When a sufficient thickness has built up, the mould is turned upside down and the remaining muddy liquid is poured out. The excess ceramic around the opening of the mould is trimmed to produce a clean edge before the mould is opened and the moulding, now in its 'green' state, is removed, ready for firing.

Pressure-assisted slip casting (see p.234) is a process that is employed for larger components.

i

Volumes of production
Versatile production volumes – anything from small-scale craft batch production to factory production.

Unit price vs capital investment
Slip casting is economical for small quantities because inexpensive moulds can be made in small workshops while maintaining fairly low unit prices. However, in industrial production the plaster moulds have a limited life and need to be replaced after approximately 100 castings.

Speed
Slip casting can be summarised by saying that 'time equals thickness'. Because of the number of operations and drying times involved, even as an industrial process slip casting still has one foot in the craft tradition, with a fair degree of labour involved.

Surface
Slip casting is a great process for achieving surface patterns on objects (such as raised flower patterns). As with all ceramic products, glazing is required.

Types/complexity of shape
Shapes can range from small and simple to large and complex, and can include parts with undercuts. Anything from bathroom products to art objects and dinnerware can be made with this process.

Scale
Large moulds can become very heavy and, given the massive amount of slip that would be needed to fill the void, slip casting may not be suited to large shapes. There is also a need for a kiln that is large enough to fire the finished product. Products such as tableware represent the average size.

Tolerances
It is hard to achieve high tolerances because the parts shrink considerably during firing and, even before that, inside the mould as the water is being drawn out of the slip.

Relevant materials
All types of ceramic.

Typical products
Slip casting is used to make any type of hollow product, from one-off pieces of tableware such as teapots, vases and figurines to high volumes of sanitary ware.

Similar methods
Pressure-assisted slip casting (p.234) and tape casting (a process used for making multilayered capacitors for the electronics industry, involving laying down thin sheets of ceramic-loaded polymers that are laminated with other materials).

Sustainability issues
The excess slip collected during moulding and after trimming is recycled back into the process to reduce material consumption and minimise the use of raw resources. The process is largely labour-assisted, which reduces the use of power and helps to balance out high energy consumption during the firing stage.

Further information
www.ceramfed.co.uk
www.cerameunie.net

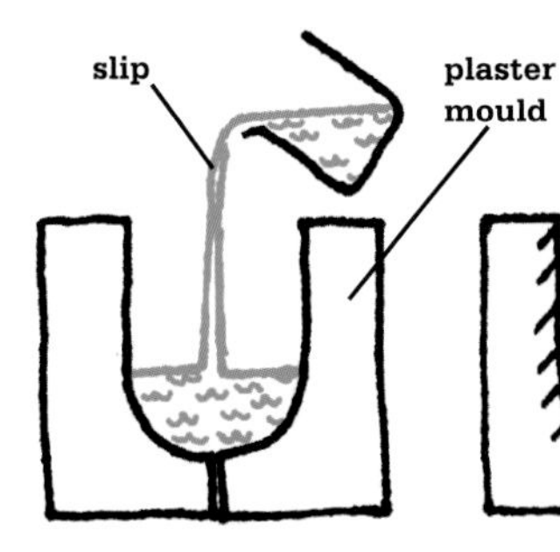

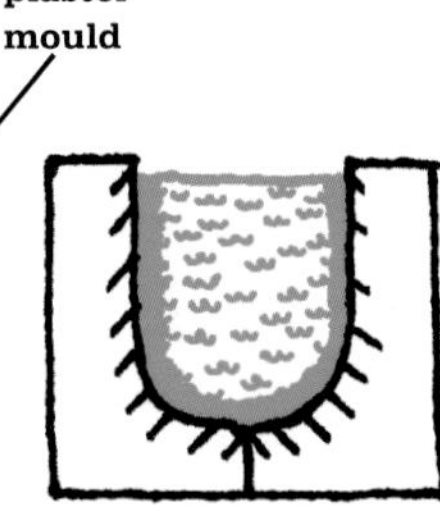

1 Slip is poured into a plaster mould, which absorbs the water leaving a layer of hard, leathery ceramic.

2 The slip is allowed to sit in the mould until a sufficient thickness has built up.

3 The mould is turned upside down and any remaining muddy liquid is poured out.

4 The excess ceramic around the opening of the mould is trimmed to produce a clean edge before the product is released for firing.

1 Empty plaster moulds.

2 Moulds filled with slip.

+

- Ideal for producing hollow ware.
- Complex forms can easily be achieved.
- Efficient use of material.
- Lends itself well to low-production runs.

−

- Labour-intensive.
- Limited control over tolerances.
- Slow production rate.
- Large-scale production requires many moulds, which themselves require storage.

Hydroforming Metal
AKA Fluid Forming

Hydroforming is a fairly new process for forming steel and other metals. It works by forcing a water and oil solution into a cylinder, or other closed shape, that is confined by a die. In essence, the process makes it possible to 'inflate' metal tubes and form metal sheets into elaborate shapes by forcing them against a die. The water pressure of up to 15,000 psi expands the material, forcing it to conform to the shape of the die to form the required component.

Tubes and cylinders are the most common starting points for hydroforming, although panel hydroforming at high pressures also exists for forming pillow shapes from two panels sealed together.

A number of benefits have resulted from this process, including parts with reduced weight and faster production times than with similar methods, such as superforming aluminium (see p.70) and inflating metal (see p.76). In order for the full potential of hydroforming to be exploited, designers need to think of it as a way of reducing costs by making something from a single material rather than having to produce a multitude of parts that need to be joined together.

Product	**t-section of a concept for a handrail system**
Designers	Amelie Bunte, Anette Ströh, André Saloga and Robert Franzheld, students at the Bauhaus University in Weimar; engineering by Kristof Zientz and Karsten Naunheim, students at Darmstadt University of Technology
Materials	hydroformed powder-coated steel; stainless steel tubes
Manufacturer	college project
Country	Germany
Date	2005

This deceptively simple, white-powder-coated steel junction from a student project for handrail systems illustrates the ability of hydroforming to create a complex form that changes from one diameter to another through a complex curve. This could otherwise only be made using conventional forming techniques that would then need to be welded together.

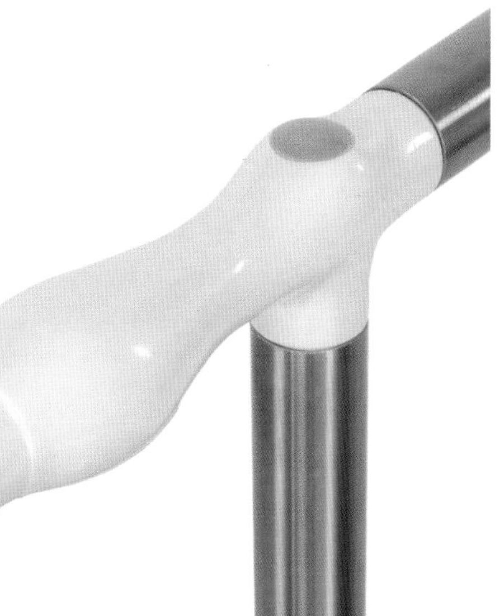

1 An example of the tooling and the die cavity into which the metal is placed.

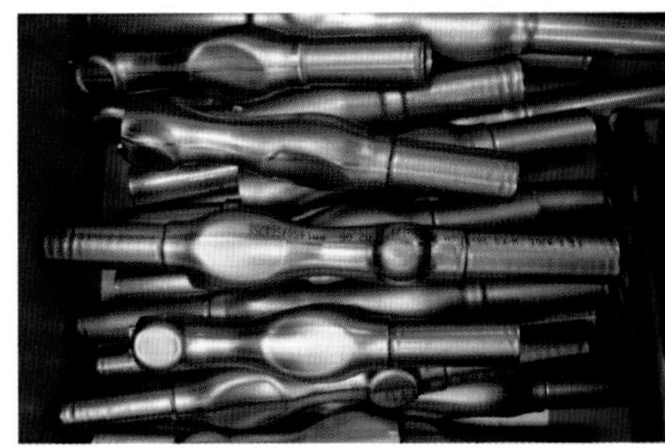

2 Semi-finished hydroformed components.

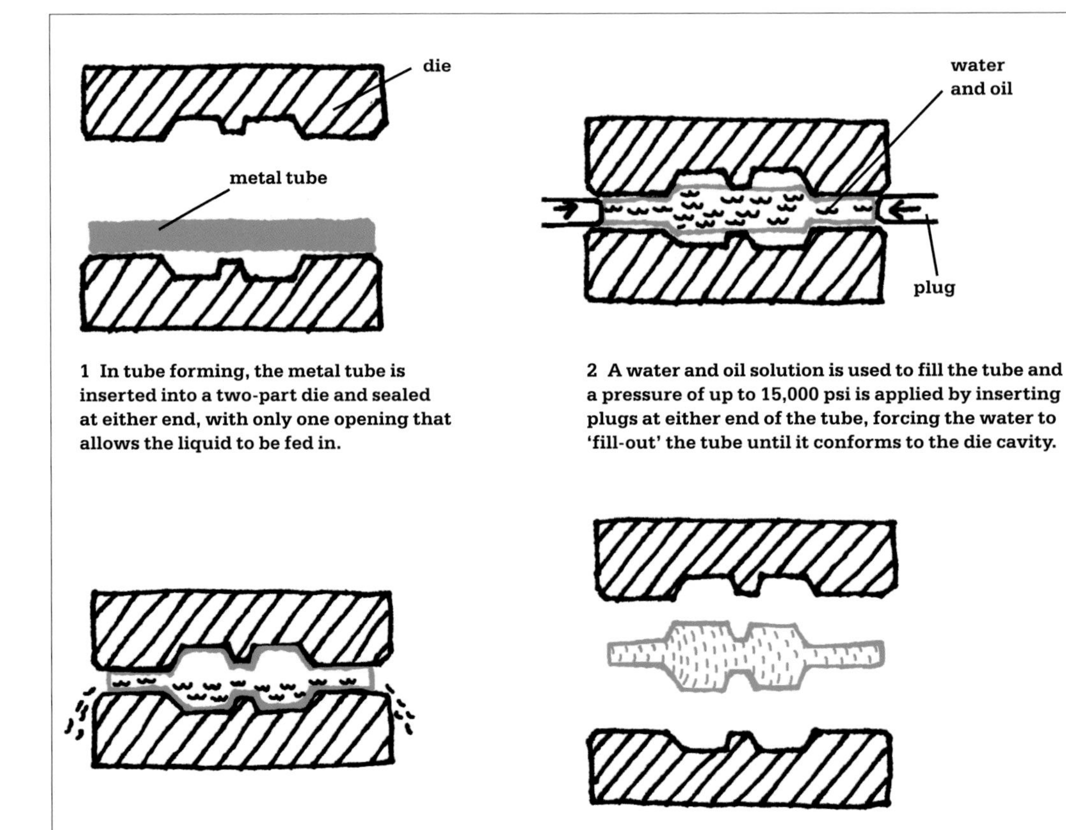

1 In tube forming, the metal tube is inserted into a two-part die and sealed at either end, with only one opening that allows the liquid to be fed in.

2 A water and oil solution is used to fill the tube and a pressure of up to 15,000 psi is applied by inserting plugs at either end of the tube, forcing the water to 'fill-out' the tube until it conforms to the die cavity.

3 The solution is emptied from the filled-out tube.

4 The final, hollow part is removed.

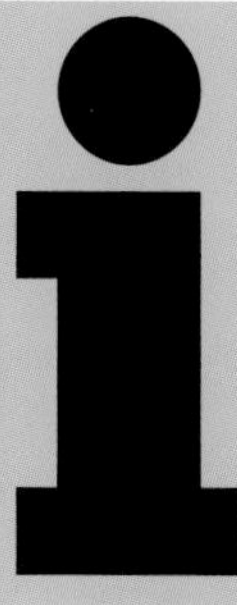

Volumes of production
High production volumes.
Unit price vs capital investment
Considerable investment in tooling required, but being able to produce components as single parts, rather than as multiple parts joined together, should help to reduce the price per unit.
Speed
In a highly automated factory setting, it is possible to achieve production cycle times of 20–30 seconds for a small part, even in a workflow set-up where parts are positioned and connected inside the die.
Surface
In general, hydroforming does not have much of an effect on the surfaces of materials. It does, however, leave small scratches and marks at the ends of a work piece from the clamps that seal the ends, but these are normally trimmed off.
Types/complexity of shape
Tubular materials can be made to bulge into quite elaborate forms. Examples of this include t-sections, which would otherwise need to be made by joining multiple components.
Scale
The bigger the part, the more pressure is needed for forming, which in turn requires a heavier mould to contain the powerful forces involved in this process. Some large car parts, including bonnets, can be made with hydroforming, although it would be difficult to manage pieces that were much larger than this.
Tolerances
Because of the die, the process allows the part to be controlled during forming to prevent wrinkling or tearing.
Relevant materials
Any metal with reasonably elastic properties that can take the high levels of tension involved, including high-grade steel and heat-treatable aluminium.
Typical products
Bicycle frames, bellows, t-sections and a variety of structural automobile components, including floor pans, van body sides and roof panels.
Similar methods
Inflating metal (p.76) and superforming aluminium (p.70).
Sustainability issues
The hydroforming technique can allow for thinner-walled parts and eliminate the need for complex joins, which can substantially reduce material consumption and weight without compromising strength and rigidity. As the metal flows instead of stretching during forming, it is less likely to work-harden, which eliminates the need for further processing such as annealing, which would require additional resources and energy.
Further information
www.hydroforming.net
http://salzgitter.westsachsen.de

- **Strong, often complex, single components due to the elimination of joints.**
- **Potential for parts with lower weight but with high strength.**
- **Ability to reduce multiple components and joins into one complex part.**

–

- **High tooling investment needed.**
- **Limited number of companies offering the process.**

Backward Impact Extrusion
AKA Indirect Extrusion

Impact extrusion is a cold process for forming metals that marries forging (see p.187) with extrusion (see p.96). In a nutshell, backward impact extrusion is a method of forming hollow metal parts by striking a metal billet (or disc), which is confined within a cylindrical or square die, so hard that the metal is forced upwards into the space between the 'hammer' (or punch) and the die. The gap between the punch and the inside of the die determines the wall thickness of the final component.

There are in fact two types of impact extrusion, forward extrusion and backward extrusion. Backward (or indirect) extrusion is used to make hollow shapes, because the punch is solid and thrusts the material around itself into the space between itself and the die.

Product	Sigg drinks bottle
Materials	aluminium
Manufacturer	Sigg
Country	Switzerland
Date	range brought to market 1998

This cut-away of the famous Sigg container shows the thin walls and the typical shapes that are a feature of impact extrusion.

The other sort of impact extrusion, forward (or direct) extrusion, can only produce solid sections. In this instance the space between the punch and the die is too small to allow metal to wrap itself around the punch. Instead, the metal is hammered downwards into a die forming a straightforward solid shape. Nevertheless, these processes can also be combined in a single operation, where the repeated action of the punch pushes material both upwards (to form a hollow top) and downwards (to make a solid, shaped base).

Designs that require outward tapering may need some post forming after extrusion, and any threaded sections, such as the bottle neck, are also added after forming.

1 The aluminium billet, which is placed on the die.

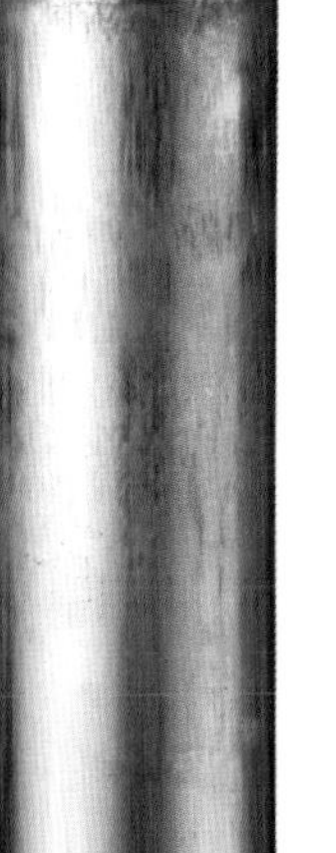

2 The cylinder created by the action of the punch using backward extrusion.

3 Tapering is added by a secondary process.

4 With the threading at the neck added, this is now recognisable as a Sigg bottle.

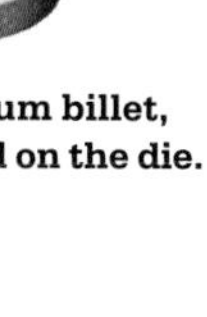

- Produces cost-effective shells in a variety of square and cylindrical cross-sectional shapes.
- Removes the issue of joints by producing components with a uniform, seamless wall.
- Inexpensive tooling compared with other high-volume processes.

- The final component is limited in length to the length of the punch needed to strike the billet.
- Only suited to parts where the length of the part is greater than four times the diameter.
- Post forming is required to add tapers or threads.
- Subject to the limitations of the die.

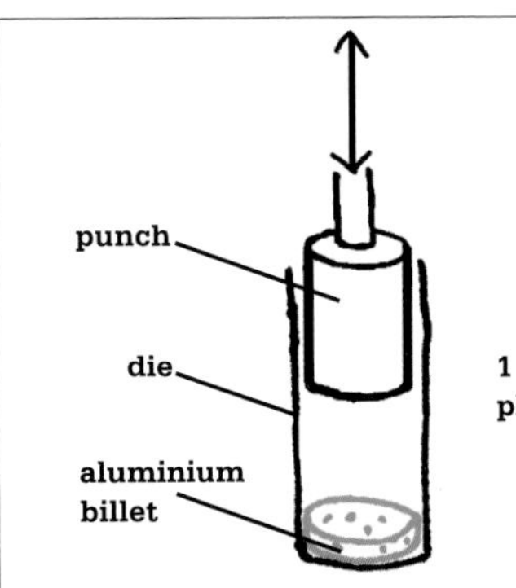

1 An aluminium billet is placed in the die.

2 The die is punched with the impact forcing the material upwards into the space between the 'hammer' and the cylinder.

Volumes of production
Impact extrusion is a high-volume production method. Depending on the size of the component, minimum quantities range from 3,000 upwards.

Unit price vs capital investment
Surprisingly, tooling is not as expensive as you might expect for a process that is used for high volumes, but the speed with which it turns out products means that it requires a large minimum order. Unit costs are very low.

Speed
The famous one-litre Sigg drinks bottles (pictured) are made at a rate of 28 per minute.

Surface
Offers a reasonably high degree of surface finish.

Types/complexity of shape
It is possible to produce thin- or thick-walled containers using backward impact extrusion, either cylindrical or square, that are closed at one end. (The forward process produces solid sections from solid rods of different shapes and sizes.) Both methods are best suited to symmetrical shapes. There are also certain guidelines regarding the ideal proportion of length and width, but you should consult your manufacturer, as these will depend on the material being used.

Scale
Suitable for parts weighing from a few grams up to approximately 1 kilogram.

Tolerances
High degrees of tolerance achievable by backward impact extrusion. (Obviously, forward impact extrusion offers greater tolerances because the final object is solid.)

Relevant materials
Aluminium, magnesium, zinc, lead, copper and low-alloy steels.

Typical products
Backward extrusion is a popular method for forming drinks and food cans, aerosol cans and similar containers. Forward and backward extrusion are used together to make such items as ratchet heads.

Similar methods
Forging (p.187) and extrusion (p.96).

Sustainability issues
Backward impact extrusion gives the metal improved strength and rigidity after forming, to allow for thinner wall thicknesses which can help to minimise material use. It is a cold-working process that requires only one single impact to form the metal into shape so energy consumption is fairly low for a process with such a fast cycle rate. In terms of material use, it is worth noting that aluminium is widely recycled.

Further information
www.mpma.org.uk
www.sigg.ch
www.aluminium.org

Moulding Paper Pulp
including rough pulp moulding and thermoforming

Paper is one of the most efficiently collected and recycled materials of the modern age. Much of what is collected is converted into pulp to make new products for a variety of industries, though these are usually simple sheets or packaging. However, it is the moulding of paper pulp using highly unusual mass-production technology that makes it particularly noteworthy.

The manufacture of moulded paper products is based on two methods: the conventional rough (or industrial) pulp process and a thermoforming process. Both methods begin by soaking the collected paper in water

Product	disposable urine bottle
Materials	paper pulp
Manufacturer	Vernacare

The mesh texture, which is subtly visible on this image, is a testament to how the water was squeezed out through a wire mesh to compact the paper pulp into a finished product. The parting lines on the mould are also visible, and the text on the product shows how the process can achieve a decent standard of surface embossing.

in a giant tank, with the proportions of paper and water based on the level of consistency needed to achieve the particular end product (typically, the amount of paper can be as low as 1 per cent). The resulting grey mixture is churned with a blade to produce the moulding compound of 'paper mush'.

Unlike most other material moulding methods, which involve the mould being stationary, the aluminium or plastic female moulds used in moulding paper pulp (which have draining holes all over them) are submerged in tanks of liquid paper pulp. The moulds are covered with mesh or gauze, which allows the water to drain out, hence the typical mesh impression that you can see on, for example, a standard egg box. A male mould is then used to compress the pulp, and a vacuum draws the water out of the mould, sucking the fibres firmly into the mould. At this point the whole thing is dried, thus forming the final product.

As well as using heat, as its name suggests, the thermoforming process involves the use of transfers and presses. After moulding, the component is picked up by a transfer, which is the negative shape of the component, and carried to a heated press that forms the final shape. It offers several advantages, including better quality surface finish but is more costly to set up.

+

- Uses recycled and recyclable material.
- Produces lightweight parts.

−

- Requires large production volumes.
- Only suitable for use with a limited range of materials.

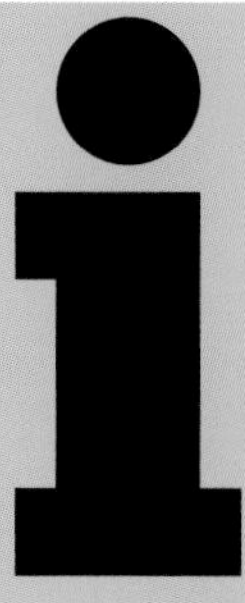

Volumes of production
Due to the high cost of tooling and the speed at which parts can be made, both rough pulp moulding and thermoforming require high volumes of production. Minimum runs of two days (about 50,000 pieces) are generally required.

Unit price vs capital investment
Tooling costs and set-up times are high. The two methods have different tooling requirements; the thermoforming method costs approximately twice as much as the raw method.

Speed
Thickness, and the amount of paper that needs to be dried, determine the speed. As a guide, the moulded inserts for four mobile-phone boxes take about a minute to produce. This is based on a multiple impression, meaning four components are moulded at the same time. These four different moulds can therefore produce 960 units per hour.

Surface
Just think of a paper egg box to get a sense of that uniquely soft, warm, biscuity surface. The rough pulp process produces one rough side, picking up an impression from the wire mesh, and a smooth surface created by the polished aluminium, or plastic, face of the mould.

Types/complexity of shape
Some fairly complex patterns can be moulded, but large draw angles need to be allowed for; forget any complex three-dimensional detailing.

Scale
Standard production allows for up to 1,500 by 400-millimetre areas – however, some manufacturers can sustain sizes up to 2.4 metres long.

Tolerances
Tolerances vary depending on the specific process. Tolerances of ± 0.5–1 millimetre are achievable using the thermoforming method. For the rough pulp process, ± 2–3 millimetres is achievable.

Relevant materials
The raw materials come from two main sources: newsprint and cardboard. The choice of material depends on the final product and the strength that is required. For strong packaging that needs to satisfy drop-test requirements (used, for example, for mobile phones, PDAs and cameras), the long fibres found in cardboard provide the best solution.

Typical products
Conventional rough pulp is used to make wine packs and industrial packaging. The thermoforming process is used to produce more sophisticated products such as mobile-phone packaging.

Similar methods
None.

Sustainability issues
Pulp is made from recycled paper products so the process helps to reduce waste and the use of raw materials in the first instance, while the pulp is recyclable at the end of its use. The conventional rough forming of the material requires little energy whereas the thermoforming process involves heat, which significantly increases energy consumption. The main drawback is the amount of water the processes require.

Further information
www.huhtamaki.com
www.mouldedpaper.com
www.paperpulpsolutions.co.uk
www.vaccari.co.uk
www.vernacare.co.uk

Contact Moulding

including hand lay-up and spray lay-up moulding, vacuum-bag and pressure-bag forming

Contact moulding is a method of forming composites by taking plastic reinforcement fibres, layering them, then applying liquid resin over the top to create a hard shell. In its simplest form – the traditional hand lay-up method – the reinforcements are laid over a mould before the liquid resin is brushed or sprayed into it. If you have ever repaired a dent or hole in an old car or boat you will probably have used a simple version of this process. In industry, it is a process for producing large-scale mouldings in composite materials, and it is one of the most frequent methods of combining various types of reinforcement fibre with thermoset resins.

The open-form moulds used in hand lay-up can be made from any material, but wood, plastic or cement are the most common. The reinforcement fibres are generally glass or carbon, but other materials, including natural fibres, can be used. A resin is then applied with a brush or by spraying, before rollers are used to squash and to distribute the mixture evenly in the mould. The spray-up method is used when larger areas are involved, using short, chopped fibres that are incorporated into the resin before spraying. In both cases, the thickness of the part is controlled by the number of layers that are applied.

Vacuum-bag and pressure-bag forming are variations of the hand lay-up and spray lay-up methods for forming composites, but they give the moulding finer detail and greater strength. The procedure is similar for both variants: in the pressure-bag method, once the materials have been laid over the mould, a flexible bag made of rubber is placed over them and subjected to pressure by clamping it, which compacts the materials, squeezing the resin and reinforcement together; with the vacuum-bag method, the part is cured inside a bag from which the air has been sucked out, forcing the materials together.

+

- The use of reinforcing fibres results in high strength.
- Other performance additives, such as flame-retardants, can easily be incorporated.
- Versatile in terms of shape and size.
- Allows thick sections to be produced.

–

- Quite a labour-intensive process.
- Requires good ventilation due to the resins.
- Other composite-forming methods (such as filament winding, see p.140) offer much higher density and strength-to-weight ratios.

With vacuum-bag forming it is possible to achieve similar results to those that you find with autoclave moulding (see p.156) but without the need for a pressure chamber. Compared with the hand and spray lay-up methods, both vacuum-bag and pressure-bag forming result in higher fibre content and density because of the use of a vacuum or pressure, which also limits the amount of potentially harmful vapour to a minimum.

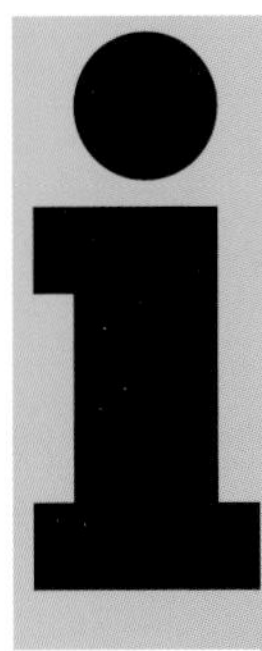

Volumes of production
The process in all these methods is always slow due to the labour involved. However, the nature of the spray-up method makes it faster than the hand lay-up process.

Unit price vs capital investment
Tooling can be inexpensive but the time taken to form parts makes them expensive to mass-produce as a high-volume process.

Speed
Depends on the type of hand lay-up technique and the size of the moulding. Spray lay-up is faster, but larger areas mean the unit speed is not always quicker.

Surface
The reverse of mouldings will have the fibrous texture of the reinforcement. Gel coats can be applied to the mould to enhance the components' surface finish. Other thermoformed skins can be applied in a secondary process for a superior surface. Vacuum-bag and pressure-bag methods allow much greater surface detail.

Types/complexity of shape
All methods are limited to open shapes with fairly thin cross-sections. Only slight undercuts are possible, depending on how far the component can be flexed when removing it from the mould.

Scale
As big as you want. Hand lay-up allows a much thicker wall thickness to be built up than spray lay-up, which reaches its maximum at about 15 millimetres. The scale of components using the bag methods is limited only by the size of the bags.

Tolerances
Due to shrinkage, tolerances are hard to control for all methods.

Relevant materials
Reinforcement materials include advanced fibres such as carbon, aramid and glass, and natural materials such as jute and cotton. Polyester is the most widely used thermosetting resin; others include epoxy, phenolic resin and silicone. Thermoplastics are far less cost-effective.

Typical products
General glass-reinforced plastic (GRP) items such as boat hulls, car panels, furniture, bath tubs, shower trays and cheap seats on the decks of small Greek ferries.

Similar methods
Transfer moulding (p.176) can achieve a similar strength. Gas-assisted injection moulding (p.201) and reaction injection moulding (p.199) can be used to create large parts, but without the strength. Other alternatives include vacuum infusion (VIP) (p.154), filament winding (p.158) and autoclave moulding (p.156).

Sustainability issues
All the processes are largely labour assisted and use a fairly low amount of energy. Use of natural fibres minimises the use of non-renewable materials. Composites are difficult to recycle at the end of their life. Yet their excellent strength and rigidity ensures a prolonged lifespan.

Further information
www.compositetek.com
www.netcomposites.com
www.compositesone.com
www.composites-by-design.com
www.fiberset.com

Vacuum Infusion Process (VIP)

The vacuum infusion process (VIP) is a method of forming composites that achieves density and strength in the end product by sucking the resin and reinforcement fibres together into a dense, solid mass. In essence, it is an advanced form of contact moulding (see p.152) and, compared with similar techniques for forming composites, it is a clean and highly effective process through which the two main ingredients can be combined in a single step.

In traditional hand lay-up methods in contact moulding, the reinforcing fibres are laid over a mould before the liquid resin is brushed or sprayed into it. In the VIP process, the dry parts of the material are stacked up over a mould. This is then covered with a flexible sheet and a seal is formed between the sheet and the mould. The air is pumped out from inside, forming a vacuum, and the liquid polymer resin is then fed into the fibres. The action of the vacuum means that the resin thoroughly impregnates the dry material, which gives the final component its density and strength.

1 A boat hull being covered in a flexible plastic sheet, ready to be sealed prior to the application of a vacuum.

2 The sheet is inspected to make sure it is completely sealed.

3 The vacuum pumps that suck the air from between the sheet and the hull.

Volumes of production
This is a slow production method that relies on the luxury of a fairly long set-up time to build the part.

Unit price vs capital investment
VIP can be used in a small workshop with basic equipment, which can be purchased from various suppliers. However, it requires a lot of trial and error, and you may suffer a high failure rate.

Speed
Slow.

Surface
Gel coats can be applied to provide the parts with a high surface finish.

Types/complexity of shape
A common application for VIP is in the manufacture of boat hulls, which should give you an idea of the level of its complexity and its scale.

Scale
The process is suited to large parts. It is difficult to make anything smaller than around 300 by 300 millimetres, because the fibre needs to be draped over or inside the mould.

Tolerances
Not the kind of process for high tolerances.

Relevant materials
As in any plastic composite method, typical resins used are polyester, vinyl ester and epoxy, combined with reinforcements such as fibreglass, aramids and graphite.

Typical products
Propellers, marine components and equipment such as a stretcher used in rescue operations, which features an aluminium frame overmoulded with a vacuum-infused composite.

Similar methods
Contact moulding (p.152), transfer moulding (p.176) and autoclave moulding (p.156).

Sustainability issues
This process is often carried out on a large scale, and its high error and failure rate results in increased waste, most of which cannot be reused. However, the fibres and resin are laid by hand which helps to balance out the high amount of energy required to heat the resin and power the vacuum. Furthermore, the vacuum ensures that only the minimum amount of resin is introduced to the fibres; any excess is sucked out, which reduces material consumption and increases the strength of the product.

Further information
www.resininfusion.com
www.reichhold.com
www.epoxi.com

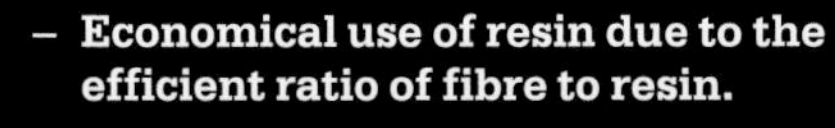

+

- **Economical use of resin due to the efficient ratio of fibre to resin.**
- **Clean.**
- **Eliminates air pockets.**
- **Higher strength-to-weight ratios than contact moulding (see p.152).**

−

- **Complicated set-up.**
- **High degree of trial and error.**
- **High failure rate.**

Autoclave Moulding

Advanced composite materials have applications across a range of industries, from premium branded sports products to engineering components. These materials offer superior strength in a lightweight moulding. However, the combination of the two distinctly different ingredients (various fibres and polymer resins) in advanced composites provides manufacturers with a challenge. They must find new ways to bring these raw materials together in a cost-effective manner that is suited to industrial production. The use of heat and pressure is a very common element within manufacturing. In autoclave moulding the combination is used to compact the raw materials together to offer the highest level of strength.

Autoclave moulding is a modified form of pressure-bag forming (see contact moulding p.152) – the composite is formed in what is essentially a pressure cooker. As a result of the applied pressure, it is one of the methods of forming advanced composite components with particularly high density. The process begins with the application of reinforcing fibres and resin onto a mould, which can be achieved through a variety of methods, such as hand or spray lay-up techniques (see contact moulding, p.152). A flexible bag is then placed, a little bit like a duvet, over the surface and the whole thing is placed in an autoclave (a sealed chamber), where heat and between 50 and 200 psi of pressure are applied, forcing the bag to squeeze itself into, or around, the mould, compressing the resin and fibres together. This forces out any potential air gaps and allows for a relatively fast curing time, compared with hand or spray lay-up. It is the squeezing together of materials under pressure, with the application of heat, that gives the final component a very high density.

Volumes of production
Batch production to medium-level production.

Unit price vs capital investment
Moulds can be manufactured from a range of materials, including modelling clay, which allows reasonably cheap tooling to be produced for short-batch production.

Speed
Although the laying together of the resin and reinforcement can be automated, the process requires manual labour and the material must pass through a number of stages. The time the material spends in the autoclave can be up to 15 hours.

Surface
Gel coats are sometimes used on the surface of the mould to provide a higher quality surface finish. Without this gel, the surface would have a fibrous texture.

Types/complexity of shape
Although the process is versatile in terms of being adaptable to different shaped moulds, it is nevertheless limited to fairly simple shapes.

Scale
Part sizes are only limited by the size of the autoclave.

Tolerances
Shrinkage does occur, so tolerances are hard to control.

Relevant materials
Suited to various advanced fibres, such as carbon fibre, and thermoset polymers.

Typical products
Widely used in the aerospace industry to fabricate high strength-to-weight ratio parts for aircraft, spacecraft and missile nose cones.

Similar methods
All forms of contact moulding (p.152), the vacuum infusion process (VIP) (p.154) and filament winding (p.158).

Sustainability issues
The combination of intense heat and pressure applied over several hours during autoclave moulding results in high energy use and increased emissions. However, the use of heat improves the performance and surface quality of the material, which may prolong the life of the product and prevent it entering the waste stream. Unfortunately, composites are difficult to recycle as it is very hard to separate the combined materials.

Further information
www.netcomposites.com

- **Increased density, faster cure times and void-free mouldings compared with moulding methods that use neither heat nor pressure.**
- **Potential for moulded-in colour.**

- **Suitable only for making hollow parts that have thick, dense walls.**

Filament Winding

Imagine impregnating the thread on a cotton reel with resin and then being able to pull the wound thread off its reel to form a rigid plastic cylindrical part: this is the essence of filament winding.

In filament winding, a reinforcement fibre combined with a polymer resin is used to form strong, hollow composites. It involves a continuous length of tape or roving (in other words, fibre) that is pulled through a polymer resin bath. The sticky fibre is wound over a pre-formed mandrel in a process that is allowed to continue until the required thickness of material is built up. The shape of the mandrel determines the internal dimensions of the finished product. If the end product is likely to be used in pressurised conditions, the mandrel may be left inside the winding to add strength.

There are various forms of filament winding that differ only in the configuration of the winding. These include circumferencial winding, where the threads are wound in parallel like the cotton thread on a spool; helical winding, where the threads are wound at an angle to the spool (which gives a woven surface pattern that is instantly recognizable); and polar winding, where the threads are run almost horizontally to the axis of the spool.

Product	**spun carbon chair**
Designer	Mathias Bengtsson
Materials	carbon fibre and polymer resin
Country	UK
Date	2003

This chair is made using a helical winding technique, though the desired effect is more gappy than is usual for components made from filament winding. This highly decorative spun structure firmly establishes filament winding – a process most often associated with engineering composites – as a design application.

1 Three composite tubes being formed on a three-spindle filament-winding machine.

2 The yellow control arm feeds the resin-impregnated fibres onto the tube-shaped mandrel. (The resin bath is out of frame.)

3 The helical winding pattern of the fibres is clearly visible.

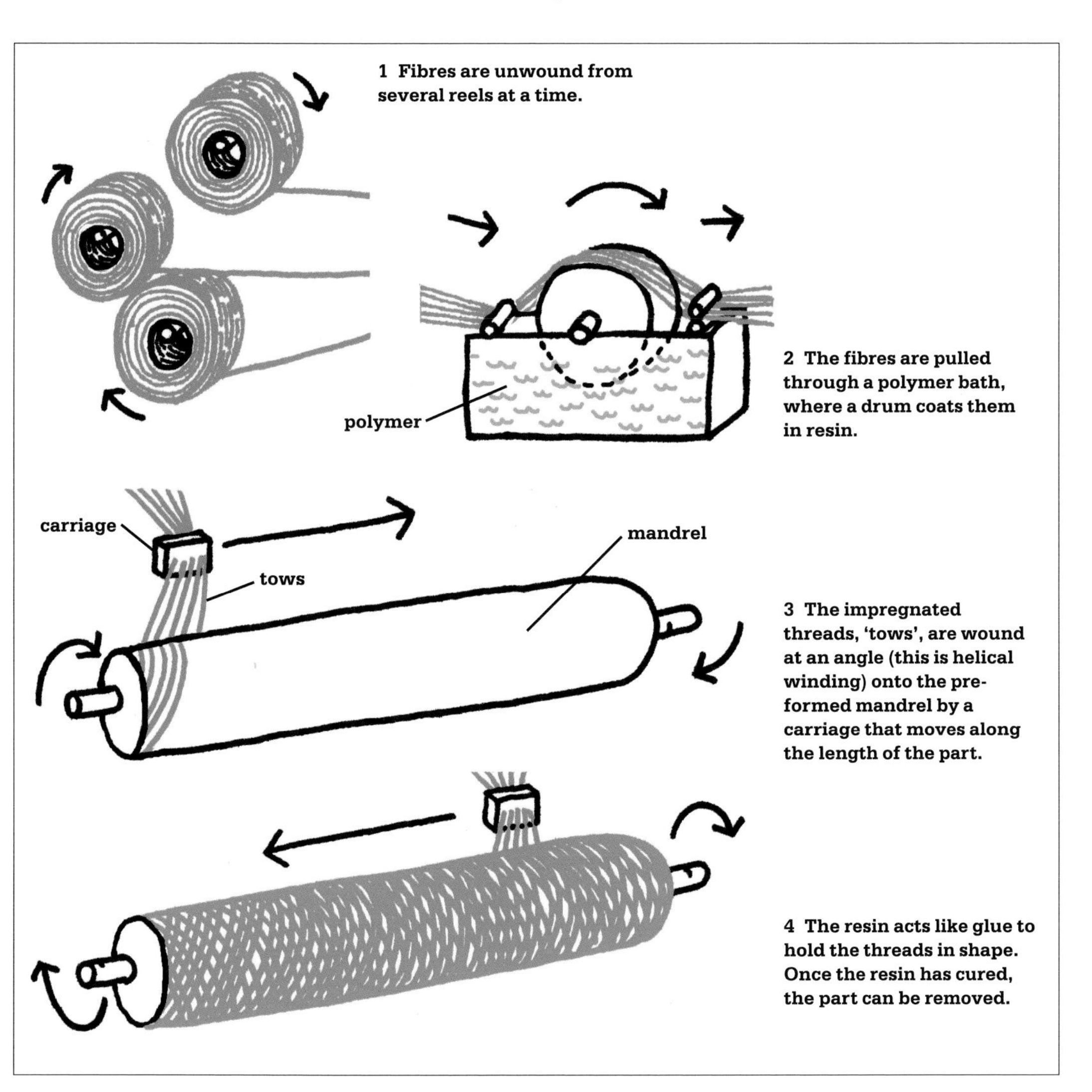

i

Volumes of production
Equally suited to one-off and high-volume production. Economical mass-production starts at approximately 5,000 units and can range up to hundreds of thousands of units.

Unit price vs capital investment
Foam tooling can be used for small runs or one-off production, as can existing aluminium bar stock, so costs can be kept down.

Speed
The speed is dependent on the shape and desired wall-thickness of the final component. However, by using a 'prepreg' system, in which the fibres are pre-coated in resin, the process eliminates the need for a resin bath. Speed is also affected by the number of 'tows' of fibre that are used, so that multiple tows result in a faster covering of the mandrel.

Surface
The internal surface depends on the finish on the mandrel, while the external surface can be finished in a number of ways, including with machining.

Types/complexity of shape
Produces very strong, thin- or thick-walled hollow components, including asymmetrical shapes.

Scale
Machines can be built to produce filament windings to a massive scale. An all-plastic, 396 metre-long motor case for a NASA rocket with a 53-metre diameter was produced in the 1960s.

Tolerances
Tolerance is controlled by the internal diameter, which is determined by the size of the mandrel.

Relevant materials
Generally used to reinforce thermosetting plastic with glass or carbon fibre.

Typical products
This is a process that is often used for closed-pressure vessels such as aeronautical components, tanks and rocket-motor housings. Because of the high strength-to-weight ratio of these parts, they are used as 'stealth' materials to replace metals in military hardware. The process is also used for its more decorative capabilities in expensive 'designer' pens made from composite materials, as well as in the chair pictured).

Similar methods
Pultrusion (p.99) and hand or spray lay-up (see contact moulding, p.152).

Sustainability issues
Filament winding is largely automated so requires electrical energy to power the motors. The high speeds at which the machines can operate help to make efficient use of this energy consumption through large volume production. The high strength-to-weight ratio is also significant and offers weight savings.

Further information
www.ctgltd.co.uk
www.vetrotexeurope.com
www.composites-proc-assoc.co.uk
www.acmanet.org

+

- **Produces components with a very high strength-to-weight ratio.**

–

- **Filament-wound components will always have a woven surface pattern unless they are post-finished.**

Centrifugal Casting

including true- and semi-centrifugal casting, and centrifuging

Centrifugal casting is a process that is based on a specific use of gravity. The same force that is at work when lettuce leaves are spun in a salad spinner, or when people are rotated in a waltzer at the funfair, is employed to thrust a heated liquid material horizontally against the inside of a mould. Once the liquid has cooled, the finished part is taken out of the mould. In industrial manufacturing, centrifugal casting is most often used to make large-scale metal cylinders that require specific surface properties within the metal component.

Centrifugal casting for metals can be broken down into three main variants: true centrifugal casting, semi-centrifugal casting and centrifuging. As you may well have guessed, each process uses a centrifugal force to throw molten metal against the inside wall of a mould to produce a variety of shapes.

True centrifugal casting is used to make pipes and tubes, and it involves molten metal being poured into a rotating cylindrical mould. The mould defines the outside surface of the final component, while the wall thickness of the final tube or pipe is determined by the amount of material that is poured in. This type of casting solves one of the problems traditionally associated with metal, because the outer surface of the component is of such a fine grain that it is resistant to atmospheric corrosion (which is a common issue with pipes), while the internal diameter is rougher, with more impurities.

In semi-centrifugal casting, either permanent or disposable moulds are employed for making symmetrical shapes such as wheels and nozzles. It involves a vertical spindle around which the mould is held, like a spinning top. It also involves a slower rotation than true centrifugal casting and parts can be 'stacked' – in other words, more than one part can be made at a time because multiple moulds can be attached to the spindle. Because the material nearest the centre (that is, nearest the spindle) rotates at a slower rate than the material furthest away, small air pockets can occur in the component.

Centrifuging is similar to semi-centrifugal casting in as much as the spinning occurs around a vertical spindle, but it is used to produce small multiple components. The metal is forced into the various mould cavities (which are only a short distance from the spindle) to produce fine details.

i

Volumes of production
From a relatively simple set-up in a jewellery workshop, to large-scale industrial production, these are processes that can be used for batch- rather than mass-production.

Unit price vs capital investment
Depends on the specific type of production: low-cost graphite moulds can be used for small production runs (up to about 60 pieces), while more expensive permanent steel moulds are used for larger runs of, perhaps, several hundred.

Speed
Slow, but it varies depending on the material that is used and the size, shape and desired wall thickness of the part.

Surface
True centrifugal casting produces an outer surface of fine-grain quality. Due to the slower rotation speed of semi-centrifugal casting, the forces in the centre of the casting are small, so gaps and porosity generally occur that need to be machined away after forming. Centrifuging enables fine details to be produced.

Types/complexity of shape
True centrifugal casting produces only tubular shapes. Semi-centrifugal casting produces parts that are axisymmetric (symmetrical around the vertical spindle) in shape only. Centrifuging is more versatile, and can produce more complex shapes.

Scale
True centrifugal casting can be used to form massive tubes up to 3 metres in diameter and 15 metres long. Wall thickness can be between 3 and 125 millimetres. Semi-centrifugal casting and centrifuging produce smaller parts.

Tolerances
The tolerances can be as good as 0.5 millimetres on the outer diameter when using metal moulds.

Relevant materials
Most materials that can be cast by other methods, including iron, carbon steels, stainless steels, bronze, brasses and alloys of aluminium, copper and nickel. Two materials can be cast simultaneously by introducing a second material during the process. Glass and plastics can also be used.

Typical products
The casting of metals is based in heavy industry, where it is used for hollow parts with large diameters. Typical parts made by true centrifugal casting are pipes for the oil and chemical industries, and water-supply components. The process is also used in the production of poles for lighting and other street furniture. Semi-centrifugal casting produces axisymmetric parts, such as storage containers for wine and milk, boilers, pressure vessels, flywheels and cylinder liners. Jewellers use centrifuging for more modestly sized metal and plastic parts.

Similar methods
Rotational moulding (p.137), although in centrifugal casting the mould is rotated at much higher speeds.

Sustainability issues
Each of the centrifugal casting techniques relies upon continuous rotation throughout each cycle, which combined with the heat required to melt the material is energy intensive. However, no waste is produced as the molten metal is added only until the required thickness is reached, keeping material consumption to a minimum. The fine finish of the outer surface attained through casting also provides metals with several years of excellent wear- and corrosion-resistance.

Further information
www.sgva.com/fabrication_processes/rna_centrif.htm
www.acipco.com
www.jtprice.fsnet.co.uk

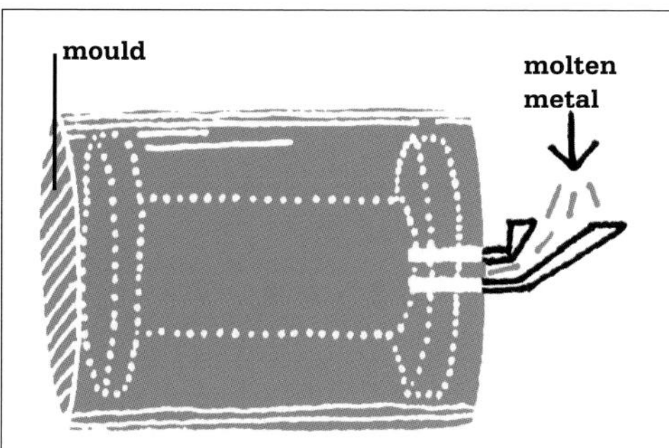

1 Molten metal is poured into a sealed mould.

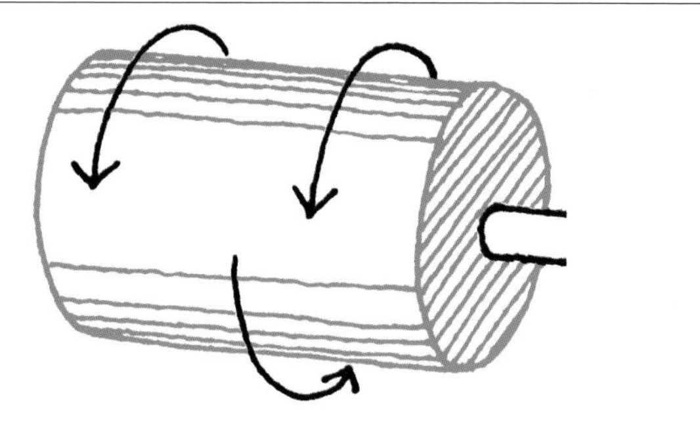

2 The mould is rotated around its axis at between 300 and 3,000 rpm.

3 The rotating action of the mould throws the metal against the inside walls of the mould. The quantity of metal determines the wall thickness of the final component.

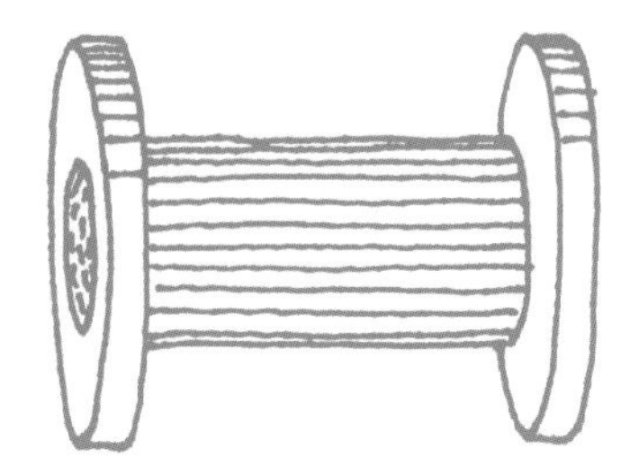

4 The finished component, removed from the mould.

+

- Parts can be produced with good mechanical properties in all directions, because the process results in non-directional grain orientation.
- The strength of centrifugal castings is close to that of wrought metal.
- With true centrifugal casting, the outer surface has a fine grain, which makes it more resistant to corrosion.
- Can achieve economical production over short runs.

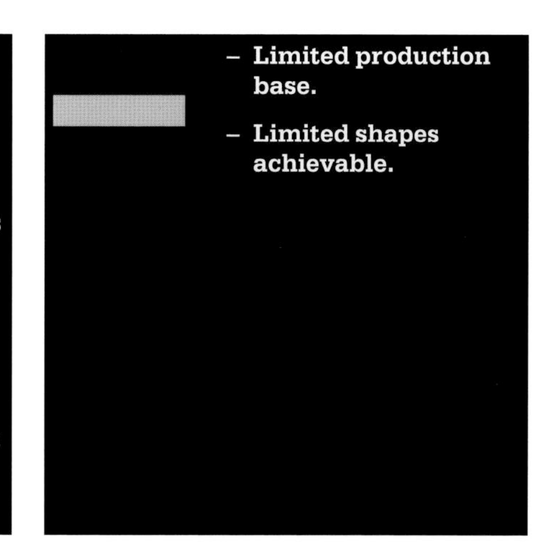

−

- Limited production base.
- Limited shapes achievable.

Electroforming

This process has changed very little since the early nineteenth century, when simple electroplating – a way of plating metals from their salts – was developed out of the initial work of British scientist Sir Humphry Davy on passing currents through electrolytes. Situated somewhere between a surface coating and a form-making technique, electroforming is a fairly unusual process. It is perhaps best explained by way of comparing it to growing a 'skin' over a shape.

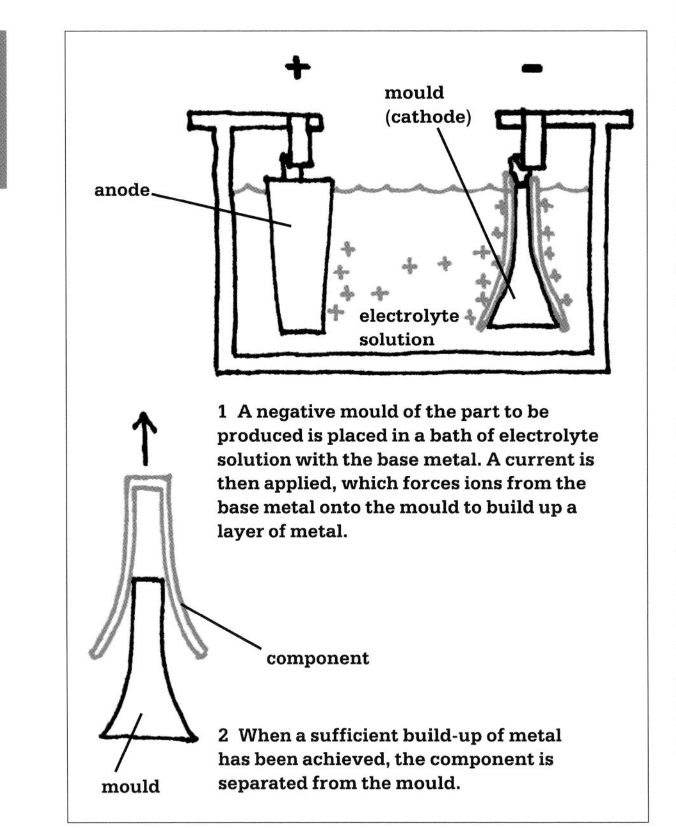

1 A negative mould of the part to be produced is placed in a bath of electrolyte solution with the base metal. A current is then applied, which forces ions from the base metal onto the mould to build up a layer of metal.

2 When a sufficient build-up of metal has been achieved, the component is separated from the mould.

The 'skin' ultimately becomes the final component once it is lifted off the mould. Essentially, it is a step on from electroplating, in which the layer of metal acts only as a coating for the original shape.

Electroforming is based on the electro-depositing of metal onto moulds. The shape that would, in simple electroplating, be coated (the cathode), in electroforming becomes a mould onto which the metalising source (the anode) is grown, in a solution of electrolyte. An electrical current forces metal ions from the anode onto the cathode. Once a sufficient build-up of metal has been achieved – and this is where it differs radically from electroplating – the component is separated from the mould. The mould does not necessarily have to be made from metals – it can be made from any non-conductive material, which can be coated with a conductive outside layer before plating.

The usefulness of electroforming lies in the fact that intricate flat and three-dimensional patterns can be easily reproduced without the need for expensive tooling, because the detail is created on the mould. The process is unique in that it creates a uniformly thin layer of material around the mould, unlike press forming (see metal cutting p.59) and sheet-metal forming (see p.50), which stretches the metal and, in so doing, leaves it an uneven thickness.

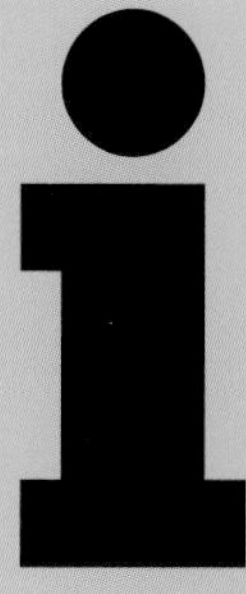

Volumes of production

Due to the length of time it takes to load moulds into the tank and produce the build-up of metal, this is not a process for high-volume or rapid production.

Unit price vs capital investment

This is an economical way of reproducing designs that are intricately patterned without needing a large investment in tooling. The cost of electroforming is partly determined by the amount of metal used, so the final unit price will depend on the surface area of the mould and the thickness of the deposited metal.

Speed

Slow, but depends on the amount of metal to be deposited.

Surface

Due to the nature of this process (the fact that it uses a mould and parts are built up gradually from tiny ions), the surface pattern can be highly intricate.

Types/complexity of shape

An ideal process for making multiple units of complex, highly decorative shapes. Making the mould from materials such as wax, which can be melted out after electroforming, means undercuts are possible.

Scale

The only limitation is the size of the electrolyte bath that holds the mould.

Tolerances

Unlike other metal-forming techniques, electroforming can produce extremely high tolerances, where the build-up of material is exactly the same anywhere on the part. This is unlike when a piece of metal is bent, a process that creates thick areas of material in corners.

Relevant materials

Nickel, gold, copper, alloys such as nickel-cobalt and other electroplateable alloys.

Typical products

A great deal of highly decorated, hollow Victorian silver tableware was produced using the technique. Today, it is still used for highly detailed silverware, but it is also used for technical laboratory apparatus and in musical instruments – a French horn, for example.

Similar methods

Simple electroplating, and as part of the micro-moulding with electroforming process (see p.250).

Sustainability issues

One of the major concerns with electroforming lies in the use of toxic substances in the electrolyte solution. However, systems have been introduced in which a special cleaning process removes any chemicals and metals from the water, which enables it to be recycled back into the process to reduce waste. Despite this, electroforming is still energy intensive as it based upon the use of a continuous electrical charge and so has relatively slow production rates.

Further information

www.aesf.org
www.drc.com
www.ajtuckco.com
www.finishing.com
www.precisionmicro.com

- **Excellent definition in detailing.**
- **Generates a uniform thickness of metal.**
- **Low tooling costs.**
- **An easy way to replicate existing products.**
- **High tolerance.**

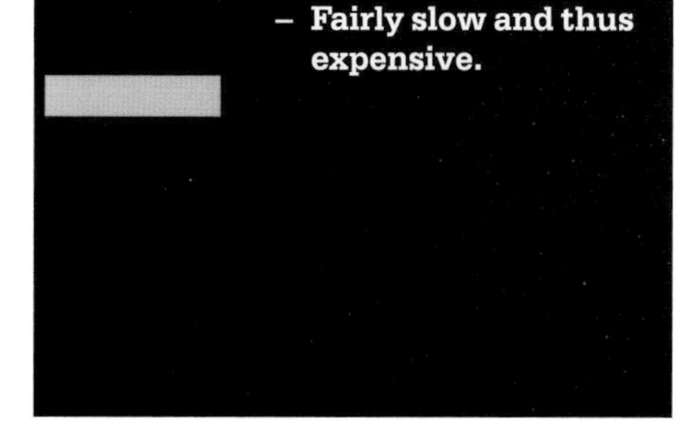

- **Fairly slow and thus expensive.**

5: Into Solid

The transformation of a material into a solid state

This chapter deals principally with a group of processes that fall within the realm of 'powder metallurgy'. This term no longer adequately describes the advanced and wide-ranging technology that exists, nor the materials. The advanced materials that are used are not always in powder form, and they include ceramics and plastics as well as metals. Processes that in the simplest terms were based on compacting metal powders into shape and then sintering the 'green' component to fuse the tiny particles together, can now be applied to many different (though mostly particulate) materials. The one exception to the powder metallurgy classification is forging, which involves transformation of an object from one solid state to another.

Sintering

including pressure-less, pressure and spark sintering, die-pressing and sintering

Sintering (a derivative of the word 'cinder') was traditionally associated with the manufacture of ceramic objects. The term is now, however, also widely used in the much larger manufacturing area of powder metallurgy. Essentially, sintering involves heating a particulate material to just below its melting point until the particles fuse together.

Various forms of sintering exist in the metals, plastics, glass and ceramics industries. Pressure-less sintering involves a powder being placed in a mould that is heated and vibrated, and then sintered. Pressure sintering involves powder being placed in a mould, vibrated and then heated, with pressure applied either mechanically or hydraulically. In spark sintering, a pulsed current passes through the mould into the powder, generating heat internally (in contrast to the above methods, where heat is applied). Die-pressing and sintering are used predominantly for ceramic or metal powders. In this process, the powder is first die-pressed into a 'green' state of the required form. This is then heated so that the particles sinter, or, in other words, fuse together.

Sintering is used to achieve high density in parts made from materials with high melting points, such as tungsten and Teflon where low porosity is needed. One of the characteristics of sintered parts, however, is that the porosity of the final component can be controlled, especially with certain materials. The porosity of some materials even after sintering can have its advantages: bronze, for example, is

+

- Suited to components with varying wall thicknesses.
- Efficient use of materials.
- Capable of forming materials that are difficult to deal with in other ways, especially very hard or brittle materials.
- Parts have good non-directional properties.
- Can produce complex forms.

−

- Requires a number of different stages.
- Difficult to achieve high tolerances due to the decrease in overall volume in sintered parts.

often used as a material for bearings, since its porosity allows lubricants to flow through. An alternative method, which eliminates porosity, is hot isostatic pressing (HIP) (see p.170).

There is also an advanced form, selective laser sintering (SLS) (see p.252), in which the application of heat is highly controlled. This method is used for rapid prototyping.

Volumes of production
Can be used for fairly low production volumes as well as for metal injection-moulded parts (see p.216), which require a minimum of 10,000 units.

Unit price vs capital investment
Tooling costs range from low to high, depending on the specific process. The nature of the process also makes it highly efficient because there is no wasted material.

Speed
This varies considerably depending on the material and the method used. For example, once compacted into shape, in the pressure-less method parts are put onto a continuous-belt furnace. Bronze typically needs 5 to 10 minutes at the centre of the furnace to sinter, while steel needs a minimum of 30 minutes.

Surface
Although the finished parts can be porous, visually there is no difference in finish compared with, for example, a standard high-pressure die-casting (see p.219) or metal-injection moulding. There is also a range of finishes that can be used on sintered parts, including electroplating, oil and chemical blackening, and varnishing.

Types/complexity of shape
Not suited to thin-walled sections. Shapes must not have undercuts.

Scale
Scale is limited to the size of the compacting press up to a maximum of 700 by 580 by 380 millimetres. Larger presses can produce approximately 2,000 tonnes of pressure, with parts requiring 50 tonnes per square inch.

Tolerances
Due to problems with shrinkage (there is a reduction in volume because of the increase of density as material flows into voids), high tolerances are generally hard to obtain unless a part goes through a secondary pressing and compaction.

Relevant materials
A variety of ceramics, glass, metals and plastic can be sintered.

Typical products
One of the most interesting examples is the production of bearings, where the natural porosity produced by the process allows lubricants to flow through the actual bearings. Other common examples include hand tools, surgical tools, orthodontic brackets and golf clubs.

Similar methods
Hot isostatic pressing (HIP) (p.170) and cold isostatic pressing (CIP) (p.172).

Sustainability issues
Sintering involves several stages of production, including intensive heating as the materials used have high melting points. This significantly increases energy consumption. However, the process allows for recovered waste materials such as iron to be reprocessed with excellent end results.

Further information
www.mpif.org
www.cisp.psu.edu

Hot Isostatic Pressing (HIP)

Product	knife from the Kyotop range
Designer	Yoshiyuki Matsui
Materials	zirconia ceramic
Manufacturer	Kyocera
Country	Japan
Date	2000

The ceramic blade of this quality knife retains its sharpness well, and has the added benefit that ceramic does not impart any taste onto food. The visible pattern, known as the 'Sandgarden effect', is lasered onto the ceramic as a secondary process.

Hot isostatic pressing (HIP) is one of the main processes for forming materials that fall under the umbrella term 'powder metallurgy' (a term that now also refers to other particulate materials, including ceramics and plastics). Heat and pressure, typically in the form of argon or nitrogen gas, are applied to powder resulting in parts with no porosity and high density, without the need for sintering (see p.168). The word 'isostatic' indicates that pressure is applied equally from all sides.

+

- **Produces parts of high density with no porosity.**
- **Because the process produces a uniform pressure, the microstructure of the final components is uniform, without weak areas.**
- **Capable of producing larger parts than is possible with other powder-metallurgy processes.**
- **Suitable for producing complex shapes.**
- **Provides an efficient use of material.**
- **Improves toughness and cracking resistance in advanced ceramics.**
- **Eliminates sintering (see p.168), which is a secondary process in other powder metallurgy-based methods of production.**

–

- **Costly set-up.**
- **Shrinkage can be problematic.**

Essentially, the process involves powdered materials being placed inside a container which is subjected to high temperature and vacuum pressure to remove air and moisture from the powder. The resulting highly compacted component is uniformly and 100 per cent dense.

The process can be used either to form components from powder or to consolidate existing components. In the latter case, there is no need for a mould as the shape has already been formed. HIP is often used for castings that need to be made denser by eliminating porosity.

Volumes of production
HIP is generally suitable only for medium-scale production quantities, typically less than 10,000 pieces.

Unit price vs capital investment
The process requires large set-up costs with expensive components.

Speed
Slow.

Surface
It is possible to achieve very high surface quality with ceramics, but other materials may require subsequent machining and polishing.

Types/complexity of shape
Simple to complex shapes are possible.

Scale
HIP caters for a range of sizes, from components measured in millimetres to large-scale products up to several metres in length.

Tolerances
Low.

Relevant materials
Most materials can be used, including plastics, but the ones that are employed most commonly are advanced ceramics and metal powders such as titanium, various steels and beryllium.

Typical products
The cost of the operation limits its use to high-spec components that require high physical and mechanical properties, such as turbine-engine components and orthopaedic implants. In advanced ceramics, HIP is used to form zirconia knife blades, silicon nitride ball bearings and oil-well drilling bits made from tungsten carbide.

Similar methods
Cold isostatic pressing (CIP) (p.172). There is also a sort of injection moulding that is suitable for ceramics.

Sustainability issues
Microshrinkage can occur during solidification, which can weaken the part internally and render it faulty, resulting in wastage. However, the defective parts can be salvaged and recycled back into the process to minimise material consumption and the use of raw resources. Moreover, the process causes the materials to strengthen and densify, which means that wall thicknesses can be reduced further to minimise material use.

Further information
www.mpif.org
www.ceramics.org
www.aiphip.com
www.bodycote.com
http://hip.bodycote.com

Cold Isostatic Pressing (CIP)

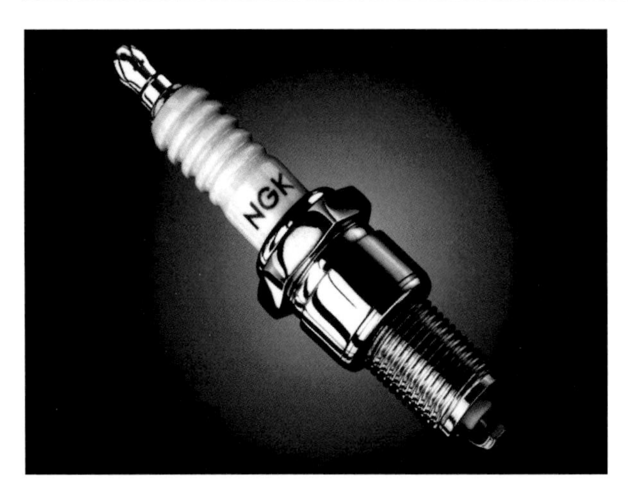

Product	spark plug
Materials	alumina ceramic
Manufacturer	NGK

The spark plug is a common product but it is made using a little-known process. The white alumina is the part that has been made using CIP.

The best way to sum up this process is to think of squeezing wet sand between your hands so that most of the water is forced out, leaving a fairly hard lump that resembles the inside of your hand. Although this sort of pressing can be done at elevated temperatures, cold isostatic pressing (CIP) is a method of forming ceramics or metal components at ambient temperatures from powders, and it involves the powder being placed in a flexible rubber bag, which squeezes around the mould when equal pressure is applied from all directions, compressing and compacting the powders into a uniform density. The process provides a uniformity of compaction around the entire component, unlike conventional forms of pressing, such as compression moulding (see p.174), which require two-part moulds.

The process is broken down into two types: wet bag and dry bag. In wet-bag pressing the rubber mould is placed inside a liquid, which, as you would expect, transmits the pressure from all directions. In dry-bag pressing, the pressure is exerted from fluid which is pumped through channels in the tooling.

Wet-bag method

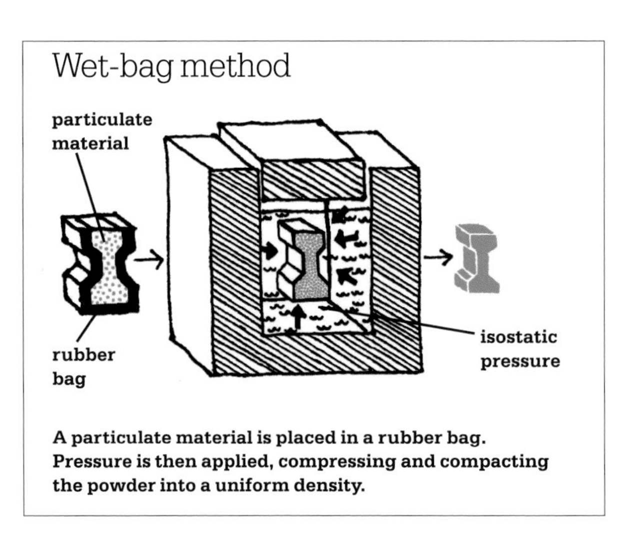

A particulate material is placed in a rubber bag. Pressure is then applied, compressing and compacting the powder into a uniform density.

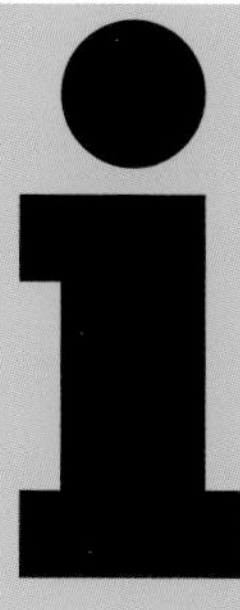

Volumes of production
Dry-bag presses are typically automated from the powder-filling to the part-removal stage, but this is a low-volume production method used to produce parts in their thousands rather than tens of thousands.

Unit price vs capital investment
Tooling can be expensive for large production runs, although existing tooling can be customised for short-batch production.

Speed
The speed depends on the particular process – for example, in the wet-bag method the rubber mould is removed from the liquid after each cycle and refilled. The dry-bag method, however, has the bag as an integral part of the mould so it does not need to be removed, but is reused to form multiple parts.

Surface
Depends on the component. Simple forms will not need any further finishing.

Types/complexity of shape
The wet and dry methods are suited to different complexities of shapes. The wet-bag method is used for complex components because of the flexible mould, which allows for easy removal of the component. It allows complex shapes to be produced, including undercuts and re-entrant angles such as collars and threads. The dry-bag presses are suited to simple shapes that can be easily removed from the moulds.

Scale
The wet-bag method is suited to large shapes, while the dry-bag presses are suited to smaller components.

Tolerances
± 0.25 millimetres or 2 per cent, whichever is greater.

Relevant materials
Advanced ceramics and other refractory materials, titanium alloys and tool steels.

Typical products
The process is suited to products that are used in harsh, aggressive environments, such as cutting tools, advanced ceramic components including carbides and refractory components. Other applications for pressed ceramics and metals include automotive cylinder-liners for aircraft and marine gas turbine components, corrosion-resistant components for petrochemical equipment and nuclear reactors, and medical implants. However, the most common product that is made by CIP is the spark plug.

Similar methods
Hot isostatic pressing (HIP) (p.170). It is also possible to use injection moulding for ceramics.

Sustainability issues
Cold isostatic pressing consumes far less energy than its hot-forming counterpart because heat is not used in either the wet or the dry method. Its high productivity level is energy efficient and it requires minimal maintenance and replacement parts. Pressure and decompression help to decrease the forming of internal stresses and cracks, thus minimizing wastage through faulty parts.

Further information
www.dynacer.com
www.mpif.org

+

- **The main advantage of CIP over other powder metallurgy methods lies in its ability to produce parts with a uniform density, with predictable shrinkage rates on a larger scale.**

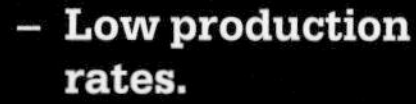

- **Low production rates.**

Compression Moulding

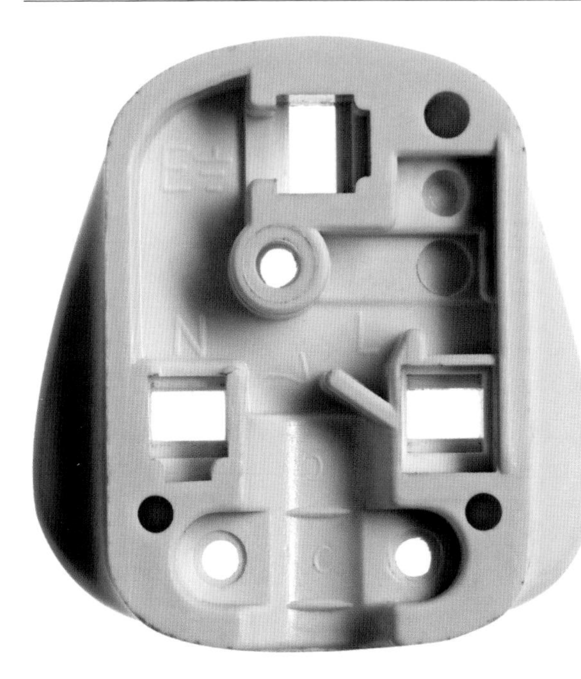

This process can be pressed into service for forming several different materials. On the one hand, it is used for producing ceramics, and on the other, it can be used to mould thermoset plastics (it was the original method for forming Bakelite), as well as fibre-based plastic composites.

To understand the basic principle of compression moulding, just think of children jamming their fists into lumps of dough to create imprints.

Product	**electrical plug**
Materials	phenol-formaldehyde plastic, also known as phenolic or Bakelite

A ubiquitous product that is an invaluable part of everyday life, but the process that lies behind it is frequently undervalued.

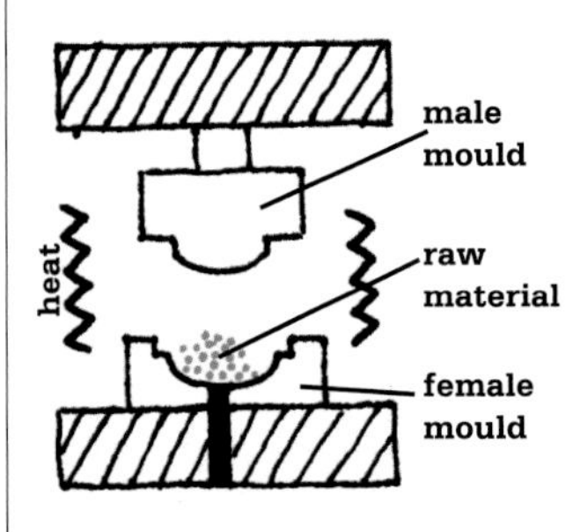

1 A two-part mould is heated and the granulated material (or sometimes a pre-form) is placed in the mould.

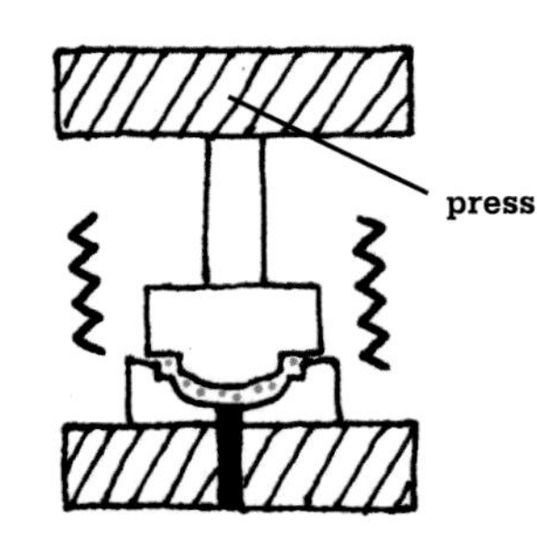

2 A press brings the lower and upper parts of the heated mould together, compressing the material into shape, the thickness of which is determined by the distance between the male and female parts.

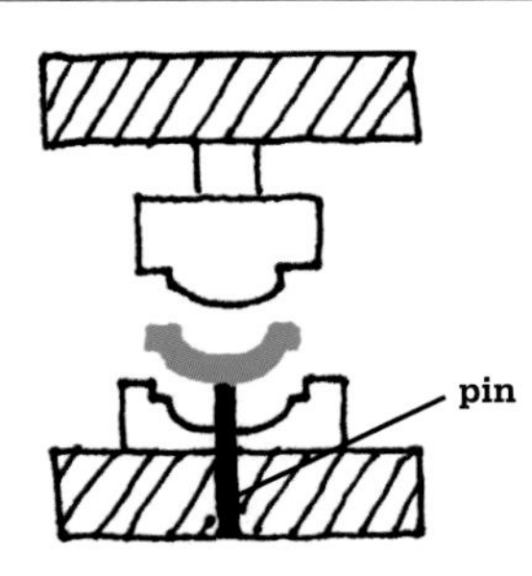

3 The moulds are separated and the formed component is ejected by pins.

The elaborated industrial process uses granules as a starting point rather than solid material, and heated moulds to replace the fist. The two-part male and female moulds can be used to process anything from thick, solid shapes to thin-walled containers.

Volumes of production
Can be equally suitable for batch or high-volume production.

Unit price vs capital investment
In relation to other plastic-moulding methods (for example, injection moulding, see p.196), tooling costs are moderate while still maintaining a low unit price.

Speed
Speed is affected by how long the mould remains closed, which is determined by part size and material.

Surface
Good surface quality.

Types/complexity of shape
Compression moulding is often used for large plastic parts with thick wall sections, which can be more economically produced with this process than with injection moulding. The nature of shaping objects with a two-part, male/female mould makes the process suitable for simple forms with no undercuts, but it also means that parts can have variable wall thicknesses.

Scale
Generally used for small parts of approximately 300 millimetres in any direction.

Tolerances
Fair.

Relevant materials
Ceramics and thermoset plastics such as melamine and phenolics, and fibre composites and cork.

Typical products
Melamine kitchenware (bowls, cups and similar products) is often made with compression moulding. Other applications include electrical housings, switches and handles.

Similar methods
Hand and spray lay-up moulding (p.152) and transfer moulding (p.176). And, although more expensive, injection moulding (p.196) could also be considered.

Sustainability issues
These depend upon the type of material used. As the process is often used for thermoset plastics, recycling parts made with this group of plastics is not an option. Wastage can be high because of the excess of material required to hold the material being formed firmly within the mould.

Further information
www.bpf.co.uk
www.corkmasters.com
www.amorimsolutions.com

- **Ideal for forming thermoset plastics.**
- **Ideal for producing parts that require large, thick-walled, solid sections.**
- **Allows for variable sections and wall thicknesses.**

–

- **Limited in terms of complexity of shapes, but good for producing flat shapes such as dinner plates.**

Transfer Moulding

Product	body panels for a London bus
Materials	glass-filled thermoset plastic

The exterior body panels on this type of bus have been made using transfer moulding. The easy flow of material through the mould cavity means that large components can be made without sacrificing control of the wall thickness.

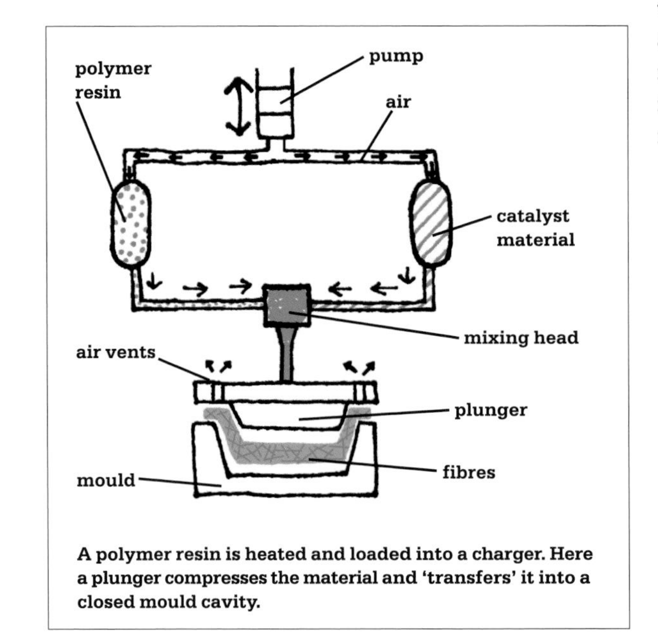

A polymer resin is heated and loaded into a charger. Here a plunger compresses the material and 'transfers' it into a closed mould cavity.

An alternative to compression moulding (see p.174) and with some of the benefits of injection moulding (see p.196), transfer moulding is typically used to make large mouldings with varying wall thicknesses and fine surface detail.

The process involves a polymer resin being heated and loaded into a charger, where a plunger compresses the material. The heated material is then 'transferred' to a closed mould cavity. The defining characteristics of transfer moulding are this heating of the material before it is transferred, and the use of a closed mould. They allow the easy flow of the material through the cavity, which results in a finer degree of control over thin-walled sections and the ability to achieve fine detail on parts. Composite materials can be made by mixing fibres with the resin, or by laying the reinforcing fibres in the mould itself.

Volumes of production
Although traditionally associated with low-volume production, recent developments have allowed transfer moulding to evolve into a full-scale industrial process.

Unit price vs capital investment
Because of the reasonably fast cycle times, transfer moulding suits high production runs, which confers the benefit of low unit costs but, as you would expect, entails high tooling costs.

Speed
This varies very much depending on the size of the part and the fibre content. Small parts can have cycle times as low as three minutes, while up to two hours is normal for large and complicated mouldings.

Surface
A good surface finish is achievable similar to that produced by injection moulding (see p.196).

Types/complexity of shape
Similar to injection moulding, but you should bear in mind that complex mouldings can increase production cycle times considerably.

Scale
It is possible to achieve a much larger scale than with, for example, injection moulding. In one recent example, the Ford Motor Company was able to swap the entire 90-piece front end of the Ford Escort for a two-piece transfer-moulded assembly.

Tolerances
Because the process involves the use of a closed mould, it achieves a greater tolerance than is possible with, for example, compression moulding (see p.174).

Relevant materials
Most often used are thermoset plastics and composites.

Typical products
Toilet seats, propeller blades and automotive components (such as the body panels for the bus, illustrated) are often made using transfer moulding.

Similar methods
Compression moulding (p.174), although it has several drawbacks compared with transfer moulding, and injection moulding (p.196), which is not as well suited to forming composites. The vacuum infusion process (VIP) (p.154) can also be used for forming composites.

Sustainability issues
Transfer moulding is capable of producing large-scale parts to provide a more efficient alternative to small-scale moulding techniques; it eliminates the need to form several different parts, and can therefore reduce material and energy use in subsequent processing. In addition, the closed mould significantly reduces styrene emissions.

Further information
www.hexcel.com
www.raytheonaircraft.com

+

- **Reasonably fast production rates.**
- **Allows complex and intricate parts to be produced.**
- **Allows large components with varying, thin- and thick-walled sections to be produced.**

–

- **Inefficient use of materials due to excess material left in runners during the moulding process.**
- **Expensive tooling.**

Foam Moulding

Unlike many other plastic-processing methods, the production of expanded plastic foam requires the material – in the case of the chair illustrated here, expanded polypropylene (EPP) – to go through a pre-expansion process before it can be manufactured. It's a bit like preparing the ingredients before you embark on a recipe.

The raw material consists of tiny beads, which, before moulding, are expanded to 40 times their original size using pentane gas and steam. This causes the beads to boil, after which they are allowed to cool and stabilise. A partial vacuum is formed inside each bead, and the beads are then stored for several hours in order for the temperature and pressure inside them to equalise. The beads are reheated and steam is used to inject them into the mould and to fuse them together. (It is also possible to perform the initial expansion of the beads within the final mould, rather than injecting the already fused beads into the mould.) The mould itself is similar to a mould that might be used in injection moulding (see p.196), with a cavity to form the final component. This recipe for moulding plastics produces materials that are up to 98 per cent air.

Enzo Mari's design for the Seggiolina POP child's chair utilises these properties in a way that celebrates the material itself. This is in contrast to

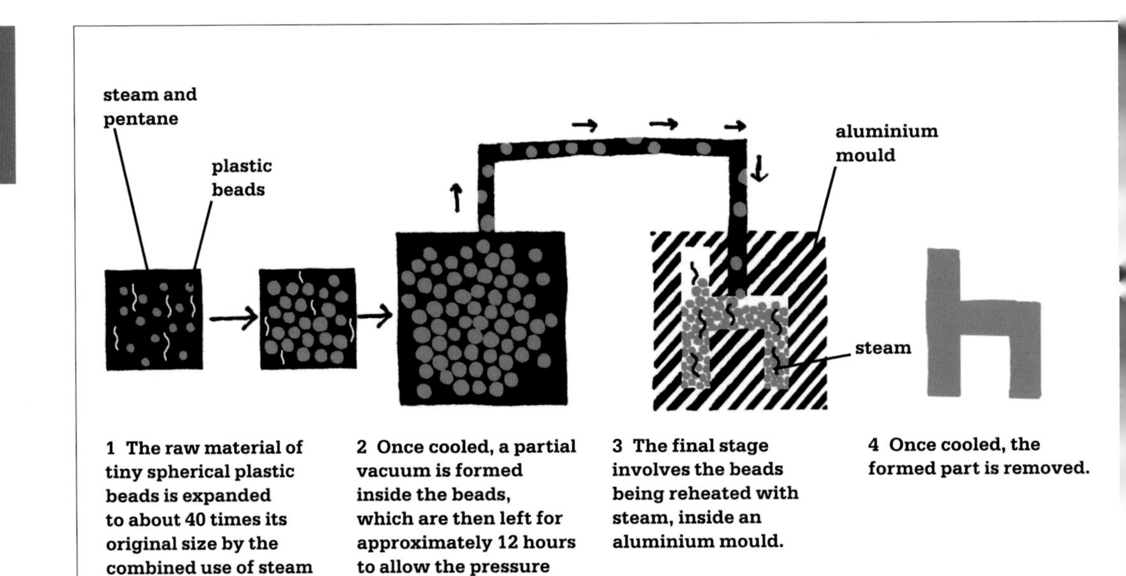

1 The raw material of tiny spherical plastic beads is expanded to about 40 times its original size by the combined use of steam and pentane.

2 Once cooled, a partial vacuum is formed inside the beads, which are then left for approximately 12 hours to allow the pressure to equalise with the external environment.

3 The final stage involves the beads being reheated with steam, inside an aluminium mould.

4 Once cooled, the formed part is removed.

its more general applications, where it tends to be hidden away in a cardboard box or under upholstery.

As well as producing stand-alone components and products, various manufacturers have developed technology that enables expanded polypropylene to be moulded directly into the casings of other components, reducing assembly times and costs.

Product	**Seggiolina POP chair**
Designer	Enzo Mari
Materials	expanded polypropylene (EPP)
Manufacturer	Magis
Country	Italy
Date	2004

The bright colours of the Seggiolina chair helps translate a traditionally industrial material and process into an intelligent, fun and lightweight product for children.

Volumes of production
High-volume production process.

Unit price vs capital investment
Aluminium tooling can be very expensive but produces highly cost-effective unit parts.

Speed
The moulding cycle times are typically 1 to 2 minutes, depending on the material.

Surface
The material can be coloured and printed with surface patterns, and graphics can be moulded into the surface. The surface is dependent on the density of the foam that you require, but all mouldings will have the textured foam finish that is typical of this type of material. It is also possible to produce different colour combinations within the same components, giving a mottled, multi-coloured effect.

Types/complexity of shape
Similar to the level of complexity that is possible with injection mouldings (see p.196), but with thicker and chunkier walls.

Scale
Foam moulding is a very versatile process that is capable of producing parts as small as 20 cubic millimetres up to blocks with a profile of 1 by 2 metres.

Tolerances
Tolerances vary a little between materials, but in general it is possible to work to an accuracy of about 2 per cent of the overall dimensions, with slightly higher figures for wall thicknesses.

Relevant materials
Expanded polystyrene (EPS), expanded polypropylene (EPP) and expanded polyethylene (EPE).

Typical products
Surfboards and bicycle helmets, packaging including fruit and vegetable trays, insulation blocks, head-impact protection in car headrests, bumper cores and steering-column fillers, as well as acoustic dampening.

Similar methods
Injection moulding (p.196) and reaction injection moulding (RIM) (p.199).

Sustainability issues
The expansive behaviour of the plastic beads reduces material use as the hollow foam structure is predominantly air. However, the extensive material preparation prior to moulding is energy intensive because of heat and pressure requirements that are repeated during shape forming. The lightness in weight is beneficial during transportation of the part, but materials are not recyclable.

Further information
www.magisdesign.com
www.tuscarora.com
www.epsmolders.org
www.besto.nl

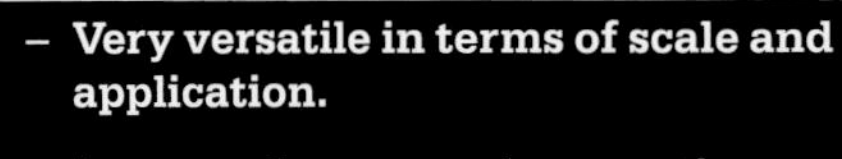

- **Very versatile in terms of scale and application.**
- **Improved structural properties.**
- **Reduced weight.**

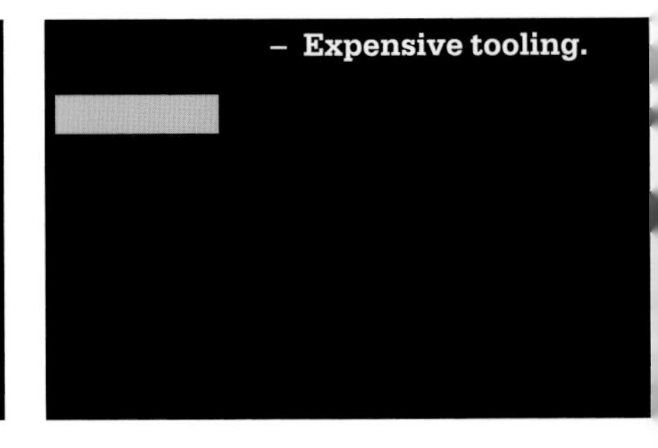

- **Expensive tooling.**

Foam Moulding into Plywood Shell

The constructional advantages and lightweight qualities of veneers were recognised by the makers of early aircraft, such as the Mosquito, and the structural use of thin veneers is nothing new in the manufacture of furniture. However, our interest in this engineered wood has now shifted to more innovative applications in furniture production.

The construction of one such example, the Laleggera chair (pictured), reveals a reverse type of tailoring. Starting with the thin veneers, the chair is constructed in the same way that a child might assemble a model kit, with a set of net shapes that are glued together at the edges, leaving a hollow shell with no structural integrity. To provide structural integrity, the shell is then injected with a polyurethane foam that, when cured, becomes rigid. This is an adaptation of the more conventional type of foam moulding (see p.178), in which foam is injected through steam into an aluminium mould where it expands to form its own skin, and can then be removed from the mould.

The great feature of the Laleggera range of furniture and the process developed by Alias is that it takes two highly uncommon materials and methods and brings them together to produce a new functional and aesthetic feature for furniture that is disarmingly lightweight.

Product	**chair from the Laleggera range**
Designer	Riccardo Blumer
Materials	polyurethane foam and wood veneers
Manufacturer	Alias
Country	Italy
Date	1996

'Laleggera' can be literally translated as 'the light one', and this chair amply deserves this description. The new manufacturing technique that made it provides an interesting marriage of materials not commonly used together.

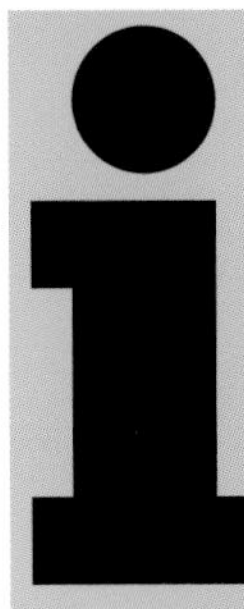

Volumes of production
This is a unique form of production so there are no points of comparison, nevertheless the manufacturers state that over 8,000 chairs were produced in 2005.

Unit price vs capital investment
This information was not made available, though it is fair to assume that a certain degree of experimentation would have been required in order that the process be set up. However, the materials (a combination of cut sheet material and injected foam) could be experimented with, in a very low-tech, cost-effective manner.

Speed
Each chair requires four weeks to produce from start to finish.

Surface
The surface finish available with this type of production is totally dependent on the plywood rather than the foam core. The surface of plywood varies depending on which type of wood is selected.

Types/complexity of shape
The shapes are determined by the ability of the plywood to be cut and assembled into a hollow shell.

Scale
The table, which is the largest piece in this collection, measures 240 by 120 by 73 centimetres.

Tolerances
Information not available, but it can be assumed that the tolerances are governed by the plywood and its ability to respond to injected foam.

Relevant materials
The chair uses a combination of plywood for the exterior structure and an internal structure of polyurethane foam.

Typical products
The unique nature of the process means that the only products manufactured are chairs and a table. However, there is no reason why the principle could not be extended to include other objects that require strength and lightweight properties.

Similar methods
This production method has been created by Alias in cooperation with the designer Riccardo Blumer. The manufacturers claim that there are no other production methods similar to this one. The nearest comparison contained in this book, though it produces a very different sort of product, is inflating wood (p.184).

Sustainability issues
The production process is heavily based on manual labour. The lightweight structure uses minimal materials with its thin veneer skin and air-filled foam. Any waste veneer can be recycled, while the injection of the polyurethane foam to fill the hollow structure creates no waste. The reduction in weight results in notable reductions in energy used during transportation.

Further information
www.aliasdesign.it

1 The frame for a table is formed from a two-part male and female press.

2 The formed components ready for assembly.

3 The structure of the table is made by gluing the formed components together.

4 Presses form the plywood around the table frame.

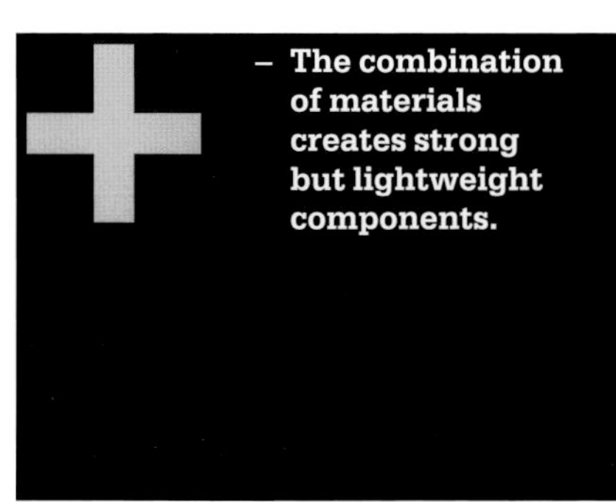

+

- **The combination of materials creates strong but lightweight components.**

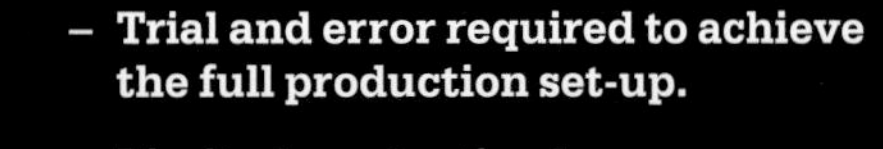

–

- **Trial and error required to achieve the full production set-up.**
- **Limited production base.**

Inflating Wood

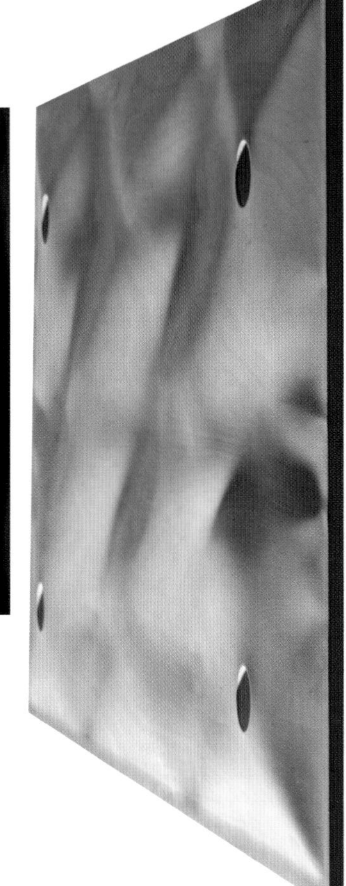

Product	door panel
Designer	Malcolm Jordan
Materials	wood veneer with foam core
Manufacturer	Curvy Composites
Country	UK
Date	2005

The undulating, compound, organic curves that are the result of this unique process allow thin wooden veneers to take on a visual quality that is perfectly suited to the natural tactile warmth of wood.

Wood may have been one of the materials used earliest by humans to produce objects, but there are many new methods of forming and transforming various forms of this basic material into new states. Most wood is transformed by being attacked with blades, but the process discussed below is a much gentler way of forming wood that relies to a large extent on the individual pattern of the grain to control the final outcome.

The process of cross-grain laminating veneers of wood was invented by the ancient Egyptians; the technique of bending plywood to introduce curves developed far more recently; and the new process of inflating wood marks the next level of sophistication in wood forming techniques. Forming wood into compound curves has always been a costly and time-consuming process, however, designer Malcolm Jordan has created a unique way of forming plywood into a series of undulating and controlled forms, though the secret of the process remains undisclosed.

The process started life as one of the many inspiring projects that have come out of the three-dimensional design course at Brighton University, on the south coast of England. Malcolm Jordan says, 'My background is in aviation. I am a licensed helicopter engineer and being surrounded by lightweight composite structures might have informed the line of experimentation. I tried a series

of experiments with various core materials between thin plywood skins.'

The final product is a composite structure, with plywood skins sandwiching a foam core. Areas of the plywood surfaces are clamped in a retaining jig. Expanding liquid foam is introduced, and the unretained surfaces move freely to form compound curves. The sizes of these wavy forms and the technique are not restricted to linear and parallel boards, but are based solely on the predetermined plywood stock sheet sizes.

Volumes of production
Most suited to batch production, rather than high-volume mass-production.

Unit price vs capital investment
Low capital investment and moderate unit cost when compared with similar methods (see below).

Speed
Dependent upon the shape or product required – for example, in the case of a batch of wall panels, the skins and frames can be pre-assembled. The foam injection is a quick process, although the assembly can remain clamped in its retaining jig for up to eight hours to cure the foam. Output would therefore be accelerated with the use of multiple retaining jigs.

Types/complexity of shape
Panels can be made either flat on one side and undulating on the other, or with two undulating surfaces that mirror each other. Because the plywood skins can be bent before the foam is injected, the process does not need to be restricted to linear or parallel boards. Solid inserts can be installed during production for 'hard points', for example to enable the attachment of legs or fittings, or to join sections together.

Scale
The scale is restricted by predetermined plywood stock sheet sizes. It has the potential for use in furniture, but there is almost certainly the possibility of a myriad of sculptural and spatial applications for architectural and interior design purposes.

Tolerances
Working with a natural material under pressure to freely form 'unnatural' three-dimensional shapes is not an exact process. Initially, there was difficulty in predicting the outcome of the curves, and the results were sometimes unexpected. However, when pressure points have been positioned and temperature and foam quantities are constant, visually similar results have been achieved.

Relevant materials
Polyurethane expanding foam (there are variations with fire-retardant additives and versions without free icocyanates). Birch-faced aero-ply ranging in thickness from 0.8 to 3 millimetres.

Similar methods
Deep three-dimensional forming in plywood (p.83) and foam moulding into plywood shells (p.181).

Sustainability issues
Plywood is manufactured from very thin slices of natural wood, a renewable and sustainable source. The expansive nature of foam, which creates hollow air-filled beads, also means that a small amount of material is consumed in relation to its overall size. The foam-injection phase limits waste as the flow is stopped when the shell is filled to the desired shape and size. However, pressure build-up sometimes causes the plywood to explode, creating an excess of material that cannot be reworked back into the process.

Further information
www.curvycomposites.co.uk

1 A metal jig is set up that will hold the plywood panel.

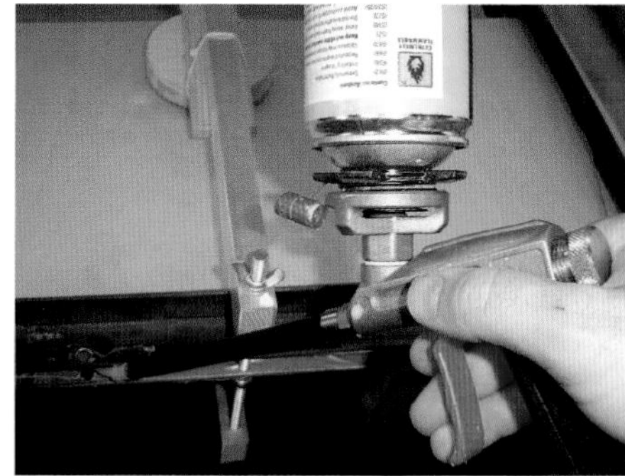

2 Final preparation of the plywood before the introduction of the foam.

3 Finishing the moulded panel.

4 A cross-section of a sample showing the foam core.

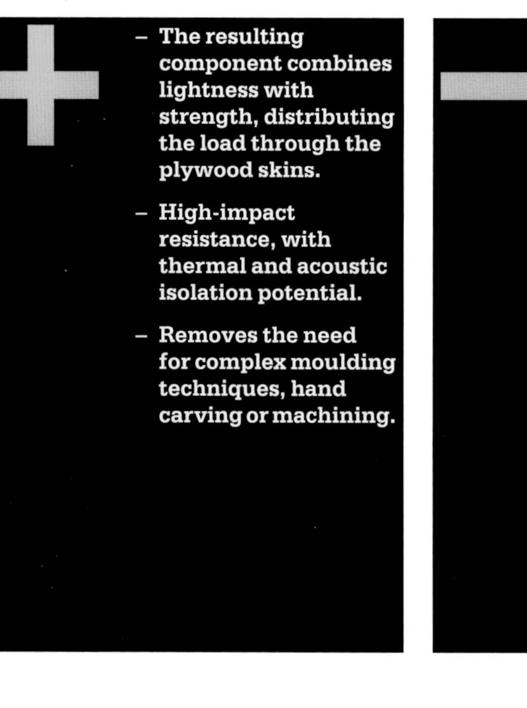

- **The resulting component combines lightness with strength, distributing the load through the plywood skins.**
- **High-impact resistance, with thermal and acoustic isolation potential.**
- **Removes the need for complex moulding techniques, hand carving or machining.**

- **Controlling the foam pressure. At present the quantity and pressure of the injected foam is regulated via inlet- and outlet-restricting devices, although experience has shown that the pressure created can be enough to rupture the plywood (with a big bang).**
- **Curving plywood can highlight flaws in the wood, often caused by manufacturing tools when the plywood's veneers were cut. Careful selection of plywood or a veneer added to the plywood overcomes this.**
- **Only available through one manufacturer.**

Forging

with open- and closed-die (drop), press and upset forging

Forging is a major process in metal forming, sometimes utilising architectural-scale machines for pounding metals into shape. It is not only a method of forming metal, it also produces a change in its physical properties, resulting in enhanced strength and ductility. In its simplest manual form – open-die forging – it involves a chunk of metal being heated to just above its recrystallisation temperature and then being formed into shape by repeated blows with a hammer, as performed by a traditional blacksmith. Movement of the work piece is the key to this method. In its more industrial incarnation, it incorporates several variations, including hot and cold forgings.

Closed-die (or drop) forging involves a very similar process to that of the open-die method described above. In this instance, however, the shaped hammer is held in a machine

Product	raw, semi-finished spanner
Materials	steel
Manufacturer	original manufacturer undisclosed, but finished by King Dick Tools in the UK
Country	Germany and UK

This unfinished ring spanner is the result of the closed-die forging process and is seen here before it is finished by drilling with a pilot hole and broached with a serrated tool to obtain the twelve points.

and repeatedly dropped onto the metal, which sits in a shaped die. The shape of the two parts determines the formed shape. Drop forging can be either hot or cold. The hot form involves the blank metal being heated, and results in stronger components due to the realignment of the grain.

Press forging involves a heated bar being slowly squeezed between two rollers, which form the metal as it is fed through. Upset forging is used for shaping the ends of the rods by compressing them as they are held in the die. Typical products produced by upset forging are nails or bolts.

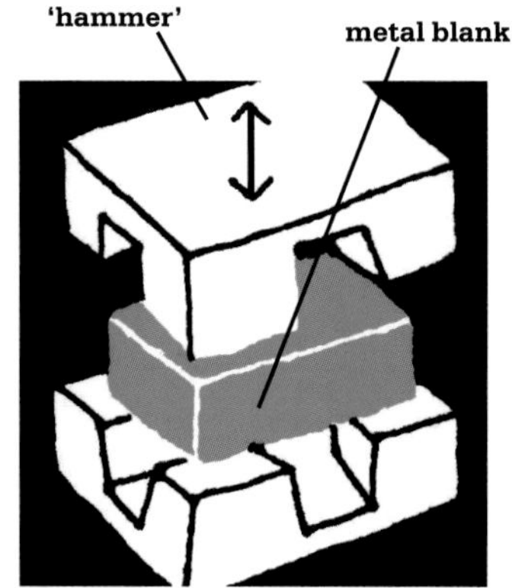

1 In hot closed-die forging, a metal blank is heated and placed in a die cavity.

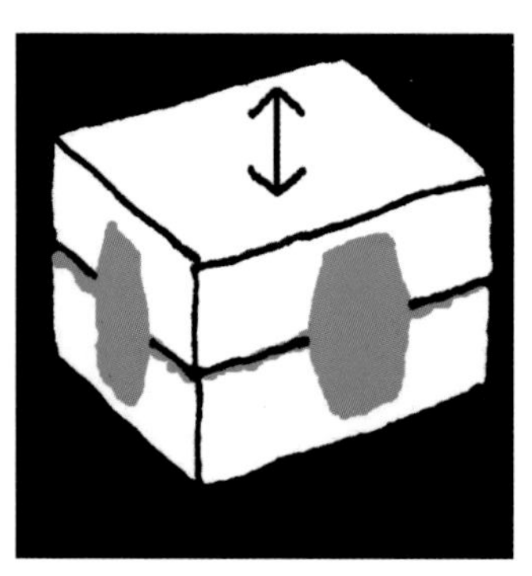

2 The male and female parts of the mould compress the metal, by means of a hammering action, into the die cavity.

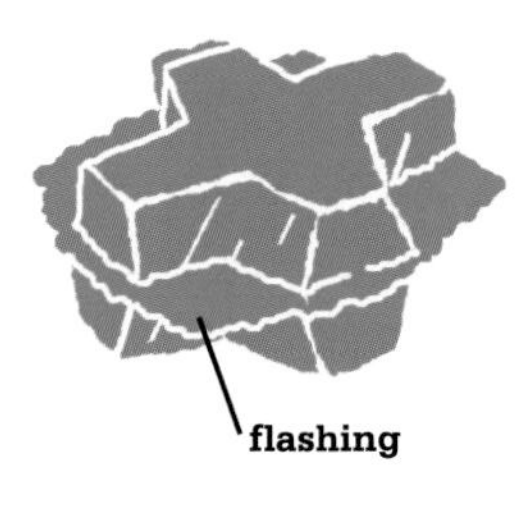

3 The part is removed from the mould ready for the flashing to be machined away.

+

- One of the main reasons for choosing forging is for the control it gives over the grain structure in the metal. It allows for the grain flow to be aligned to specific shapes, making the part stronger and more ductile.
- No gaps or voids occur in the metal, as they can in die casting (see p.219) and sand casting (see p.228).
- Less waste than with runners and sprues.

–

- Forged parts often require machining to remove the excess metal that is left when the two halves of the die are brought together.

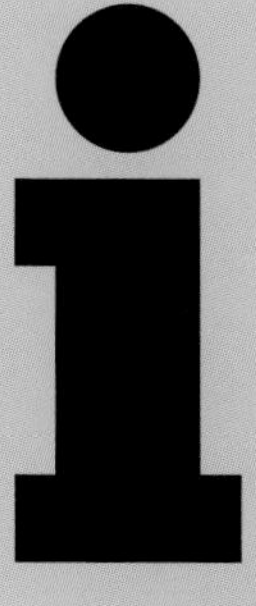

Volumes of production
From simple hand forgings up to about 10,000 units.

Unit price vs capital investment
In hot open-die forging, done by hand, the cost is based on the skilled manual labour. In automated methods, tooling costs can be very high.

Speed
Quite slow, which is partly due to the fact that 90 per cent of all forging processes are hot processes, so that the work pieces need to be heated before forming.

Surface
Forged parts will generally need to be machined in order to achieve a good, smooth surface and to remove flashing, which is the result of metal being squeezed out into a flat web around the outside of the part.

Types/complexity of shape
The type of forging process will dictate the complexity and type of shape that is possible. In drop forging, draft angles are generally required, and parting lines need to be designed in order for complex shapes to be formed. Draft angles vary and are dependent on the type of metal used.

Scale
Forging can be used for parts that weigh from a few grams to those reaching 0.5 tonnes.

Tolerances
High tolerances are difficult to achieve, partly due to the wearing of the die. Different metals offer a range of tolerances.

Relevant materials
With hot forging, most metal and alloys can be formed. However, the ease with which they can be forged varies enormously.

Typical products
Because of the increased strength of forged components (compared with cast metals), a large number are used in aircraft engines and structures. Other applications include hand tools such as hammers, wrenches and spanners, and swords – notably Samurai swords.

Similar methods
Powder forging (p.190). Impact extrusion (p.146) and rotary swaging (p.106) are both forms of forging.

Sustainability issues
The increased strength the material acquires during forging can increase the durability and lifespan of the final product. However, the heated forging techniques consume high amounts of energy, which increases emissions and subsequent effects on the environment. In addition, a significant amount of excess metal is produced, and secondary machining and further energy use is required to trim it. Fortunately this excess can be recycled.

Further information
www.forging.org
www.iiftec.co.uk
www.key-to-steel.com
www.kingdicktools.co.uk
www.britishmetalforming.com

Powder Forging
AKA Sinter Forging

Powder metal forging is a process that sits within the realm of powder metallurgy. It combines sintering (see p.168) and forging (see p.187) to produce finished parts. As in other forms of powder metallurgy, the process begins with the forming of the metal powder into a 'green' state in a die. At this stage, the component is known as a 'pre-form', and is slightly different in shape from the final component. The pre-form is sintered to obtain a solid component, which is removed from the furnace, coated with a lubricant such as graphite, and transferred to a forging press. Here, the final component is formed in a closed-die forge, which forces the metal particles to interlock and become a solid, dense mass. The extra compaction provided by this process gives a highly dense, non-porous component.

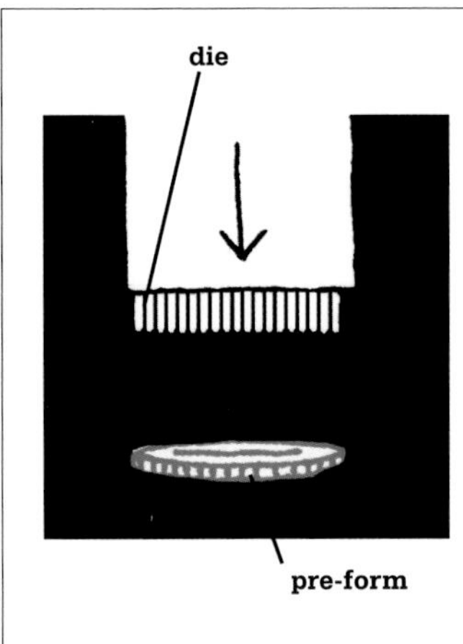

1 Metal powder is compressed into a 'green' state in a die to obtain the preform.

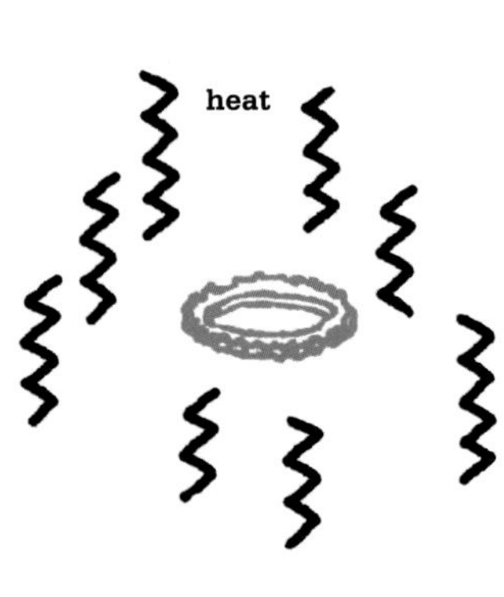

2 The pre-form is sintered to obtain a solid component, which is removed from the furnace, coated with a lubricant such as graphite, and transferred to a forging press.

3 The final component is formed in a closed-die forge, which forces the metal particles to interlock and become a solid, dense mass.

Volumes of production
High volumes, typically over 25,000 units.

Unit price vs capital investment
This high-volume production process is expensive, partly because of the need for two sets of dies. Large volumes are needed to produce economical components.

Speed
Depending on the set-up and the component size, extremely high speeds are possible to achieve.

Surface
Good surface, which does not need secondary processing such as heat treating.

Types/complexity of shape
The process is capable of producing complex shapes. Powder forging can accommodate a high degree of varying wall thicknesses, which can be as low as 1 millimetre. Undercuts are not possible.

Scale
Similar to drop forging and press forging (for both, see forging, p.187) – think of a spanner or a gear (around 200 millimetres in diameter) for reference.

Tolerances
Part of the advantage of powder forging is its ability to produce parts with higher tolerances than other forging methods.

Relevant materials
Most ferrous and non-ferrous metals. A large number of powder forgings use iron with small amounts of copper and carbon.

Typical products
Engineering components for a range of industries, including automotive parts, connecting rods, cams, hand tools and transmission components.

Similar methods
Drop forging and press forging (p.187) and compression moulding (p.174).

Sustainability issues
Powder forging offers greater precision and less excess material than conventional forging, so requires only minor secondary processing to make more efficient use of energy. It still requires high temperatures to create material flow and this has a large impact on energy consumption and emissions. In addition, over several hundred runs the intense impact pressure between the die and the substrate material can result in greater maintenance requirements.

Further information
www.mpif.org
www.gknsintermetals.com
www.ascosintering.com

+

- **No gaps or voids in the metal, which can occur in, for example, sand casting (see p.228).**
- **Compared with other powder metallurgy processes, powder forging provides parts with greater ductility and strength.**
- **Efficient use of material, with less wastage than in other forms of forging (see p.187).**
- **Requires far fewer post-forming operations than other forging methods.**

–

- **Expensive tooling that requires large volumes of production.**

Precise-Cast Prototyping (pcPRO®)

The Fraunhofer Institute in Germany is one of the world's biggest research organisations concerned with materials and manufacturing. One method of production that has recently been developed by the institute is precise-cast prototyping.

Precise-cast prototyping (or pcPRO®) is a method for rapid prototyping that combines casting and milling operations in a single machine. It is a two-stage process, with the first stage involving a milling machine (see p.20) cutting a mould into an aluminium block using information from a CAD file. This mould is filled with a polymer resin. Once the resin has hardened, the same milling machine cuts it to a precise final shape. The essence of this process is that it allows for one side of a product (the moulded side) to be replicated exactly each time the mould is filled, but the top (milled) side may be adapted according to the information contained in the CAD file.

A product prototype usually requires numerous adjustments before it is optimised, forcing the modelmaker

Product	sample components
Materials	polymer resin
Manufacturer	Fraunhofer Institute
Country	Germany
Date	2004

These sample components, shown from both the top and underside surfaces, are an example of the machined CAD-cut details. The cutting lines on the surface are visible, as is the flat cast side.

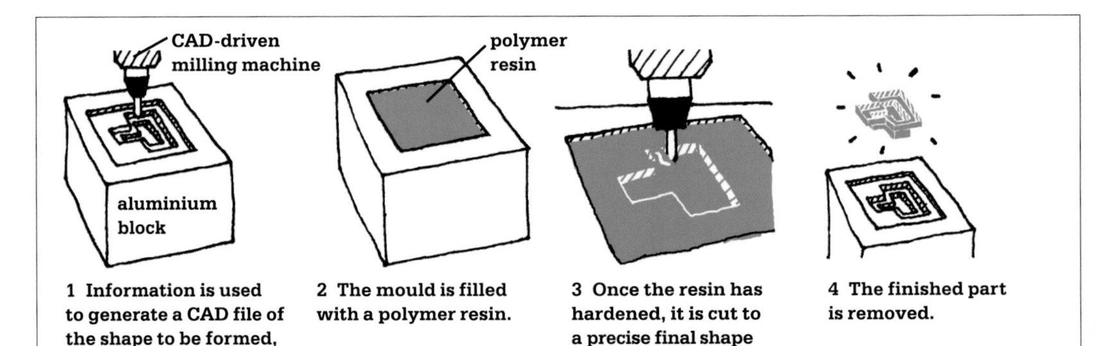

1 Information is used to generate a CAD file of the shape to be formed, which is fed into a milling machine, where the mould is cut into an aluminium block.

2 The mould is filled with a polymer resin.

3 Once the resin has hardened, it is cut to a precise final shape by the same milling machine.

4 The finished part is removed.

to start from scratch each time. With precise-cast prototyping, however, changes are only ever made in the CAD data. The main advantage is that for components such as housings for various electrical products, which have one side where the shape needs to be fine-tuned, multiples can be cast using the mould, with only one side being altered with CAD files.

Volumes of production
This is a CAD-driven process so it is suited to both one-off and batch production, though, obviously, the moulded side remains constant, so 'one-offs' only differ on their milled side.

Unit price vs capital investment
The tooling (the mould, in this case) is made using the same machine that makes the component, which means precise-cast prototyping is highly cost-effective.

Speed
Milling of the mould typically takes between half an hour and two hours; casting and curing of the resin and milling of each part takes a minimum of an hour, depending on the part's complexity.

Surface
The surface quality corresponds with the normal quality of milled surfaces.

Types/complexity of shape
The shape is limited only by the CAD drawing and the cutter (or cutters), though extremely complex shaped parts or undercuts in the inner contour can be made by five-axes milling only (that is, one cutter moving along five trajectories) and undercuts in the outer contour require special mould inserts or silicone parts.

Scale
The scale of the parts made on a standard machine is 250 by 250 by 150 millimetres.

Tolerances
Depending on the machine's accuracy, commonly some 10 microns.

Relevant material
A two-component resin.

Typical products
Complex shaped parts with high-tolerance outer surfaces and low-tolerance inner surfaces. The process is used for the rapid prototyping of bodies of mobile phones, cameras, car parts, and electric and computer accessories.

Similar methods
Conventional milling (p.20) and casting methods. Other prototyping techniques, including stereolithography (SLA) (p.246).

Sustainability issues
Moulding, creating the form and machining the part in a single process allows for incredible energy efficiency on many levels. By reducing the machinery used, energy consumption is cut dramatically, along with the emissions from transportation of parts between manufacturing locations. Alterations to the milled surface can be carried out immediately without the need for a whole new mould and test runs, significantly reducing material use.

Further information
www.fraunhofer.de

+

- **Permits the combination of automated and shape-specific manufacturing.**
- **Time- and cost-effective.**
- **High-quality finish.**

−

- **Limited number of manufacturers offer this method.**

6: Comp

Parts with complex shapes and surfaces

These processes can be described as 'plastic-state forming' because of the soft, malleable and, generally, hot state of the materials as they are moulded. It is these methods of production that are most responsible for the explosion in the number of cheap, moulded plastic products now available. Nevertheless, the payback for achieving complexity at a low cost per unit is the level of investment required for tooling. This chapter contains many of the established methods of high-volume mass-production, such as injection moulding in plastic and die-casting in metals. It also investigates methods of adding finishing materials to complex shapes.

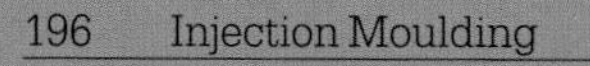

ex

Injection Moulding
with water injection technology (WIT)

Is injection moulding the mother of all plastic-processing techniques? It is through this process that we are able to transform plastic into a mass of packaging, toys and casings for electronics. It could well have been an injection mould that French philosopher Roland Barthes was referring to when he wrote, in his *Mythologies* (1957), of '. . . an ideally shaped machine, tabulated and oblong (a shape well suited to suggest the secret of an itinerary) effortlessly draws out a heap of greenish crystals, shiny and fluted dressing-room tidies. At one end, raw, telluric matter, at the other, the finished, human artefact, hardly watched over by an attendant in a cloth cap, half-god, half-robot.'

The process employs plastic pellets, which are fed from a hopper into a heated cylinder, which contains a screw. The screw carries the hot plastic, slowly melting it, and finally injecting it at high pressure into a series of gates and runners, which feed the polymer into a water-cooled steel mould. Once the part has solidified under pressure, pins eject the finished part from the mould.

Water injection technology (WIT), or water-assisted injection moulding, is a relatively new technology that promises several advantages over conventional injection moulding and gas-assisted injection moulding (see p.201). It is based on several variations, which either employ the injection of water to ram the melt (polymer) into the mould, or use water injection as a means of forcing the polymer outwards, to the walls of the mould, to create hollow parts. The use of water eliminates some of the problems that are associated with gas-assisted injection moulding, such as migration of the gas into the plastics. In addition, due to the fact that water cannot be compressed, a greater degree of pressure is produced than can be provided by gas, which results in several advantages in terms of the complexity and finish of the final parts. Faster cycle times are also achievable due to the cooling effect of the water.

Product	**BIC® Cristal® ballpoint pen**
Designer	Marcel Bich
Materials	polystyrene (shaft); polypropylene (lid and plug)
Manufacturer	BIC
Country	France
Date	1950

Millions of BIC® Cristal® ball pens are sold worldwide every day. All the elements of this iconic ballpoint pen are made using injection moulding, except for the cartridge and nib.

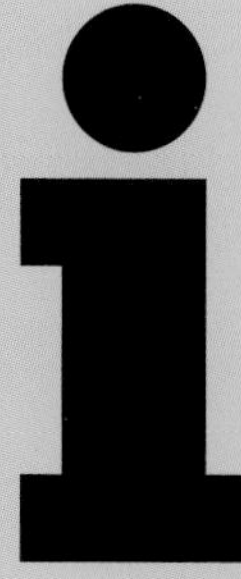

Volumes of production

Small injection-moulding manufacturers can be found to produce simple components of 5,000 units or less. However, the minimum quantity is generally accepted to be 10,000 units.

Unit price vs capital investment

Unit price is very low, but this must be set against the high tooling costs, which can run into tens of thousands of pounds.

Speed

Cycle times vary depending on the type of material, wall thickness and the geometry of the part. As an example, simple bottle tops have the fastest cycle times of between 5 and10 seconds. A common speed for more complex parts is between 30 and 40 seconds.

Surface

This is determined by the steel mould, and can vary from spark-eroded to highly glossy. The points where ejector pins are located in the mould need to be considered when designing a part, as these leave small, indented circles. Parting lines, where the various parts of the mould come together, also need to be considered.

Types/complexity of shape

If the volumes of production are particularly great, injection moulding can be used to form highly complex parts. However, features such as undercuts, variable wall thicknesses, inserts and threads will add significantly to the cost of the tooling. Generally, injection moulding is suited to thin-walled sections.

Scale

Micro-injection moulding is a specialist area and there are certain manufacturers who specialise in parts that are often less than 1 millimetre in size. For large-scale products such as garden chairs, it is worth considering gas-assisted injection moulding (p.201), and, where thick walls are required, try reaction injection moulding (RIM) (p.199).

Tolerances

±0.1 millimetre.

Relevant materials

Predominantly used for thermoplastics, but thermosets and elastomers can also be used.

Typical products

It is impossible to state 'typical' products produced by injection moulding because its use is so widespread, from sweet packaging (tic tacs™ boxes, for example) to medical implants.

Similar methods

The equivalent process for metals is metal injection moulding (MIM) (p.216) or high-pressure die-casting (p.219).

Sustainability issues

This process offers a precise, controlled and optimised use of material and energy. Water injection technology can help to improve the energy efficiency of injection moulding through faster cycle times and by forming a closed-loop cycle in which the water is not disposed of but is put back into the process. However, with its ability to churn out plastic parts quickly and cheaply, injection moulding can be accused of encouraging disposability as there is no cost incentive to reuse products. It can also come under fire for the release of toxins during heating and further emissions through high energy use.

Further information

www.bpf.co.uk
www.injection-molding-resource.org

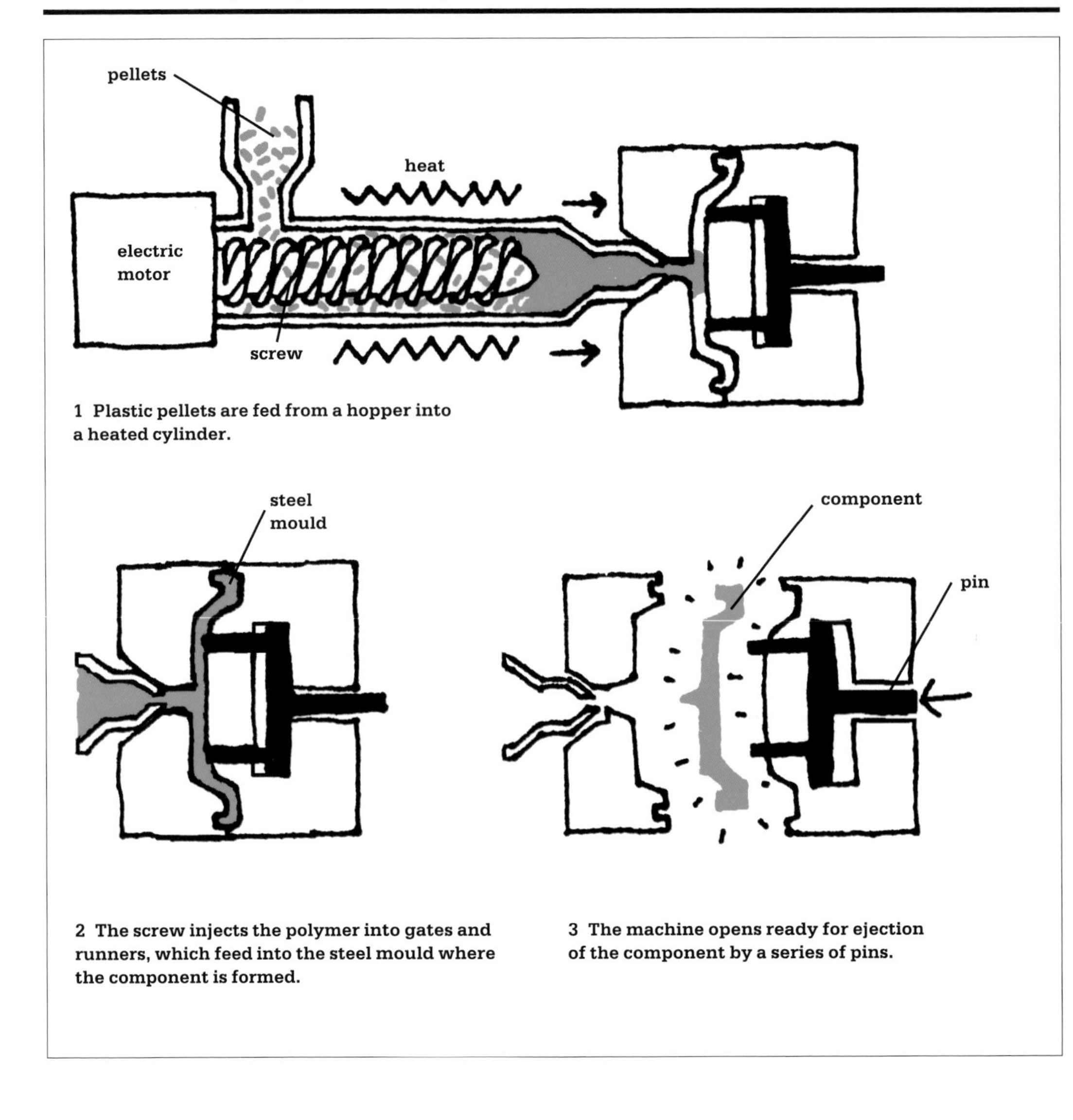

1 Plastic pellets are fed from a hopper into a heated cylinder.

2 The screw injects the polymer into gates and runners, which feed into the steel mould where the component is formed.

3 The machine opens ready for ejection of the component by a series of pins.

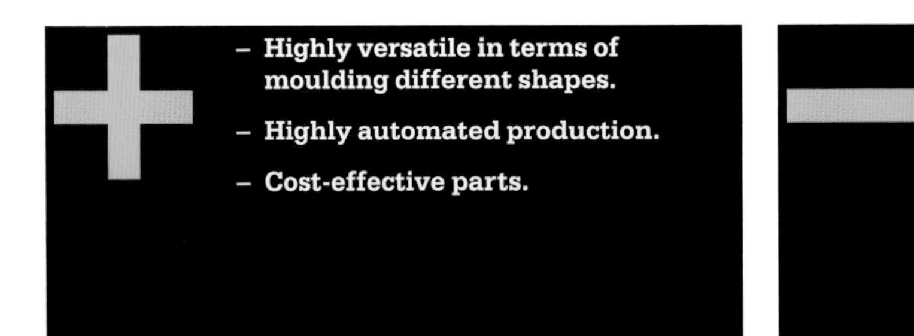

- Highly versatile in terms of moulding different shapes.
- Highly automated production.
- Cost-effective parts.

- Involves considerable investment and high production runs.
- Can involve long lead times.

Reaction Injection Moulding (RIM)
with R-RIM and S-RIM

Reaction injection moulding (RIM) is a process that is used for producing structural foam components. Unlike standard injection moulding (see p.196), which uses pellets as the starting point, RIM involves feeding two reactive thermosetting liquid resins into a mixing chamber. They are then injected through a nozzle into the mould, where an exothermic chemical reaction produces a self-forming, smooth skin over a foam core. Depending on the formulation of the resin, parts produced using RIM can either be soft foams or solid, highly rigid components.

Composites can be produced by introducing short or long fibres into the mixture, to add reinforcement. This form of production can be broken down into two categories: reinforced-reaction injection moulding (R-RIM) and structural-reaction injection moulding (S-RIM).

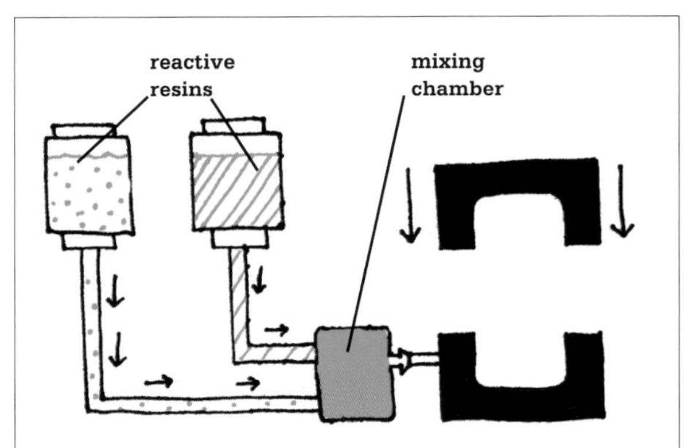

1 A combination of two reactive resins is fed into a mixing chamber.

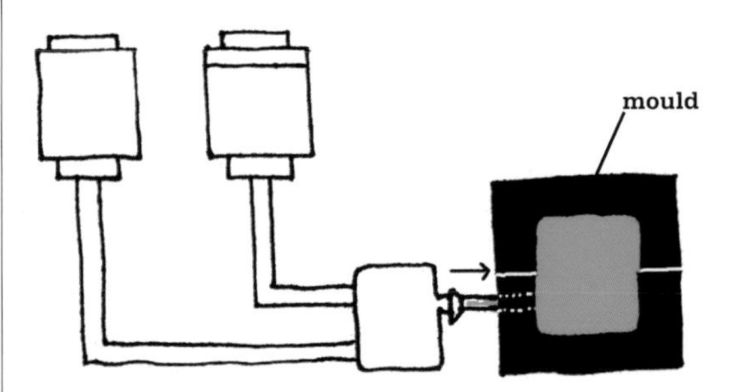

2 From this chamber the resins are fed into the mould, where an exothermic chemical reaction produces a smooth skin over the foam core of the final component.

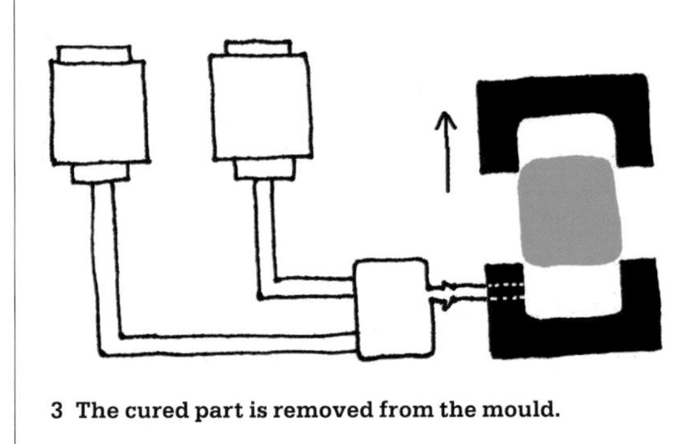

3 The cured part is removed from the mould.

Volumes of production
Suited to high-volume production, however, because of the potential to use inexpensive, low-strength moulds, low-volume production is also a realistic possibility.

Unit price vs capital investment
This is a low-pressure process with low tooling costs compared to those of standard injection moulding (see p.196). However, set-up costs are high, so large numbers of units need to be produced in order to be economical.

Speed
This is not a rapid process, unlike standard injection moulding. Cycle times are considerably longer and, depending on the size and complexity of the part, can take several minutes, as opposed to seconds, per part.

Surface
The foams produced using this process are 'self-skinning' and sometimes form a hard skin similar in quality to that formed in standard injection moulding, while retaining the foam core.

Types/complexity of shape
Large and complex solid shapes are possible, with the potential to create varying wall thicknesses in the same component. Typical wall thicknesses of RIM parts are a chunky 8 millimetres.

Scale
Suited to large-scale components up to 2 metres long.

Tolerances
High tolerances.

Relevant materials
RIM is often used to form dense polyurethane foams. Other common materials include phenolics, nylon 6, polyester and epoxies.

Typical products
Large foam mouldings, rigid and flexible alike, are manufactured using RIM for use in products such as car bumpers and trim, industrial pallets, casings for large-scale electronics and refrigerator door panels.

Similar methods
Injection moulding (p.196) and transfer moulding (p.176). Also, gas-assisted injection moulding (p.201), which allows for complex, large lightweight parts, although it is not suited to foams.

Sustainability issues
Material use is considerably lower than with conventional injection moulding as the expansive nature of the foam produces a predominantly hollow structure that better resists shrinkage while maintaining excellent strength. In addition, the heating temperatures are considerably lower, which results in reduced energy consumption and emissions. One disadvantage is that the slower cycle times mean energy use is less efficient than it is with conventional injection moulding. Materials, however, are not recyclable.

Further information
www.pmahome.org
www.rimmolding.com
www.plasticparts.org

- **Allows for varying wall thicknesses within the same part.**
- **Because of the low pressures and temperatures required for this process, tooling costs can be low compared with other high-volume plastic methods.**
- **Produces parts with a high strength-to-weight ratio.**
- **Suitable for making large parts.**

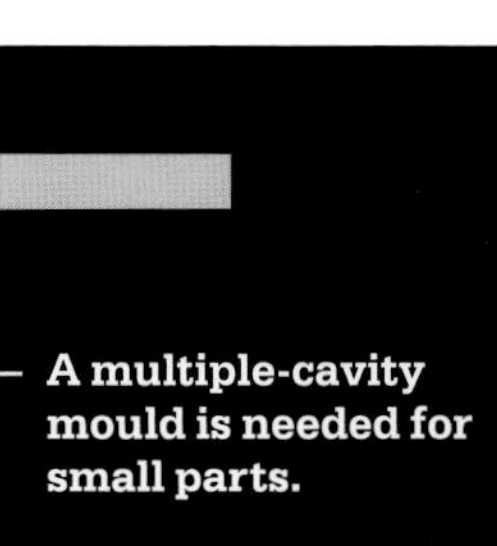

- **A multiple-cavity mould is needed for small parts.**

Gas-Assisted Injection Moulding

In standard injection moulding, thermoplastics are heated and injected into a mould (see p.196). Channels in the mould act to cool the plastic part before it is released from the mould. During cooling, the part shrinks and moves away from the walls of the mould and, to compensate for this, more material is injected into the mould.

An alternative to this widely used method is to inject gas, usually nitrogen, into the mould cavity while the plastic is still in its molten state. This internal force counteracts the shrinkage by inflating the component, forcing it to remain in contact with the surface of the mould until it solidifies, resulting in parts with hollow sections or cavities.

There are two types of gas-assisted injection moulding: internal and external moulding. The former is the most widely used, with the external method being used when greater definition in the part, or a larger surface area, is required. This is achieved by injecting a very thin layer of gas between one surface of the plastic and its adjoining mould cavity.

Exploiting the reduction in weight provided by this process, the Italian manufacturer Magis has produced a range of furniture that redefines the rules for designing large-scale plastic products. The ubiquitous low-grade garden chairs, produced by standard injection moulding, that you find in your local DIY retailer are made with a thin cross-section and have a strong, stable structure. By contrast, the Magis range, designed by Jasper Morrison, appears to be solid, but the inside is hollow.

Product	Air-Chair
Designer	Jasper Morrison
Materials	polypropylene, with glass-fibre reinforcement
Manufacturer	Magis
Country	Italy
Date	1999

This stackable chair, while sturdy enough to withstand considerable bulk, is lightweight, hollow and economical, all of which are the advantages of using gas-assisted injection moulding.

Volumes of production
Strictly a high-volume production process.

Unit price vs capital investment
Like standard injection moulding (see p.196), it combines low unit costs with high investment.

Speed
Due to the fact that material is injected only once and cools more quickly than it would in standard injection moulding, cycle times are reduced.

Surface
One of the key advantages of this form of injection moulding is the superior finish. During standard injection moulding, stress usually occurs along the flow line inside the mould, resulting in warping. The introduction of gas helps to distribute the pressure evenly and eliminate the stress and flow lines in the plastic at specific points.

Types/complexity of shape
Injection moulding is one the best methods of producing complex shapes, and gas-assisted injection moulding is no exception. Depending on how much you want to spend on tooling and the number of parts in the mould, you can achieve some highly complex shapes.

Scale
From casings for small electronic components to large pieces of furniture.

Tolerances
With greater control over the material and less shrinkage than in standard injection moulding, tolerances are higher.

Relevant materials
Most thermoplastics, including high-impact polystyrene, talc-filled polypropylene, acrylonitrile butadiene styrene (ABS), rigid PVC and nylon, and also composites.

Typical products
Virtually all mouldable parts can be made using gas-assisted injection moulding. External gas-assisted injection moulding is often used for components with large surface areas, such as car-body panels, furniture, refrigerator doors and high-end plastic garden furniture.

Similar methods
Injection moulding (p.196) and reaction injection moulding (RIM) (p.199).

Sustainability issues
Gas-assisted injection moulding can allow for significant reductions in material use as hollow, lightweight parts can be produced. Cycle times are also much faster, with significant savings in energy consumption in comparison to traditional injection moulding. In addition, the reduction in weight can be beneficial in reducing fuel consumption during transportation of the product.

Further information
www.magisdesign.com
www.gasinjection.com

+

- **Allows for components to be made with variable wall thicknesses.**
- **Reduced cycle times.**
- **Reduced weight.**
- **Less sink marking than in conventional injection moulding (see p.196).**
- **Consumes 15 per cent less energy than standard injection moulding.**

–

- **Because of the extra parameters involved – the handling of gas, regulation of pressure and cooling – potential problems need to be addressed in advance, requires experience, and often, a fairly complicated set-up.**

MuCell® Injection Moulding

The traditional injection moulding technique has been used to mass-produce plastic components for everything from mobile phone casings to shoes for over 50 years. With the evolution of new composite materials, it is no surprise that MuCell® injection moulding has evolved along with them. MuCell® is a process that can be applied to both injection moulding and extrusion, but introduces a new substance – microcellular foam – into the mix. On average the process allows for a weight reduction of ten per cent and reduced moulding cycle times of 35 per cent.

A polymer mixed with a foaming agent is forced into the mould cavity under very high pressure. Once the polymer has spread throughout the cavity, nitrogen gas is shot into the mould at a very high heat at which it reaches a point between being a vapour and a liquid. This is called the

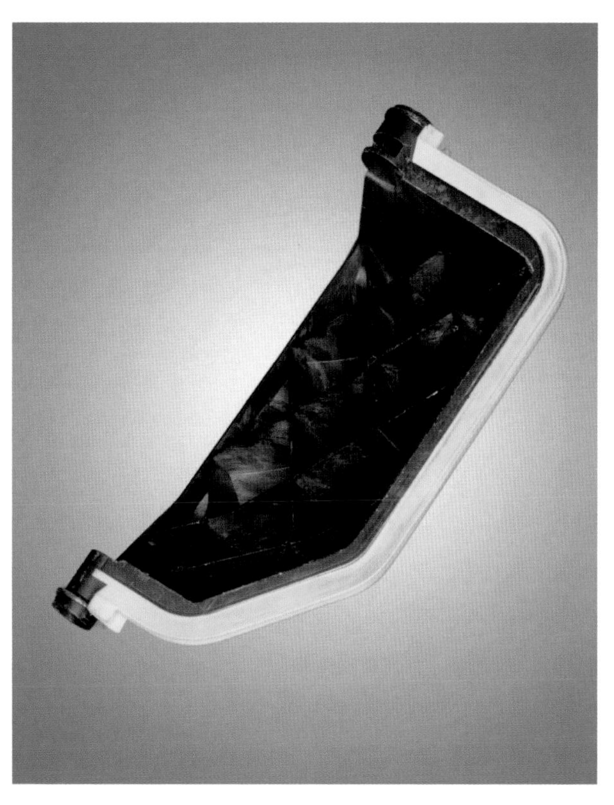

Product	HVAC valve component
Materials	Talc-filled, PP (black plastic moulded with MuCell®); second sealing gasket moulded in solid with thermoplastic elastomer
Manufacturer	Valeo
Country	Germany

MuCell® technology is used in this product to create a strong but lightweight (seven per cent) component. The material is dispersed equally which results in less shrinkage of the part.

'critical temperature' and is crucial to the behaviour of the polymer. Once the gas reaches a higher temperature than this it dissolves into the molten polymer. However, as the pressure inside the mould begins to decrease the gas changes state again, and separates to form a consistent cell structure within the polymer. The cells are microscopic in size but they have incredible strength while being very low in weight. The polymer mix has now formed a microcellular foam.

Because the structure of the microcells is consistent and uniform, the stress within the mould cavity is dispersed equally, which results in less shrinkage than there would be in traditional plastics, which do not have a uniform structure. The moulded product is significantly lighter in weight and has less viscosity than if it were conventionally moulded. There is also a closer compliance to mould shape and dimensions because of the expanding nature of the foam. The uniform structure provides the moulded product with excellent stiffness in addition to thermal and conductive insulation.

The technique is not suitable for everyday moulded products, but has been designed for engineered plastic components that require high precision and accuracy. It is a single-phase process – the polymer and gas are injected within the same cycle – and increase in material flow provides the potential for producing thinner parts. To give an idea of the scale the process works with, the parts it produces typically have a wall thickness of no more than 3 millimetres.

+

- Weight of moulded parts is significantly reduced.
- Increased dimensional stability because of the uniform cell structure.
- Cycle time is reduced due to the reduction in weight and viscosity.
- No shrinkage during cooling.

–

- Limited number of manufacturers.

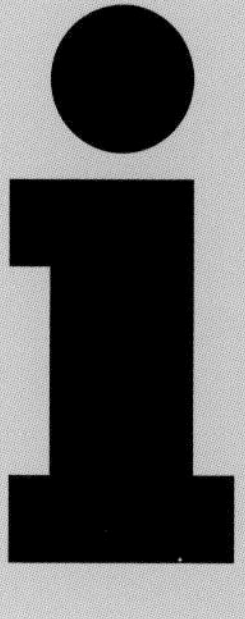

Volumes of production
The process is applicable to high-volume production runs that are comparable with injection moulding or extrusion.

Unit price vs capital investment
Reduced cycle time and usage of materials lowers the cost significantly compared with traditional injection moulding. However, the process requires investment in specific equipment that costs more than that used for conventional injection moulding.

Speed
The manufacturers maintain there is a 15 per cent to 35 per cent improvement in cycle times compared with conventional injection moulding of thermoplastics.

Surface
The use of the gas creates parts with increased flatness and with less potential to warp.

Types/complexity of shape
The process offers the ability to create finer details and thinner wall thickness than conventional injection moulding or extrusion.

Scale
The size of components that can be produced typically ranges from something the size of a latch pin weighing a few grams to a large automotive part weighing several kilograms. Typically the wall thickness of MuCell® parts is less than 3 millimetres, 2.5 millimetres for talc-filled PP.

Tolerances
$\pm$ 0.1 millimetres

Relevant materials
A range of thermoplastics, of which engineering plastics such as PA, PBT, PEEK and PET are known to perform better. Materials often perform better when filled with fillers such as glass fibres.

Typical products
Most of the key applications for this process are currently based on automotive components because of the reduced weight of the moulded parts. Applications for base plates for power tools have been implemented, where nylon filled with glass fibres replaces metals while maintaining the flatness needed in this type of application.

Similar methods
Gas-assisted injection moulding (p.201).

Sustainability issues
Material consumption is significantly reduced due to the expansion of the foam, which in turn decreases the weight of the component. The viscosity of the material is less, thus speeding up cycle times and making efficient use of energy.

Further information
www.trexel.com

Insert Moulding

Insert moulding is a branch of multi-component moulding (also called two-shot moulding), which is a method of combining different plastics in the course of only a single manufacturing process. Insert moulding refers to the stage of the process where parts (made from a variety of materials, including metal, ceramic and plastics) are inserted to increase strength in the plastic component. Injection moulding (see p.196) is the dominant element in this method of manufacture, with the inserts being placed in the mould prior to the injection of plastics.

Multi-component insert moulding using injection moulding exists in two forms. In the first method, known as rotary transfer, two materials are injected into the same mould cavity with the mould having been rotated. The second method, commonly

Product	**Stanley DynaGrip Pro' screwdriver**
Designer	Stanley in-house design
Materials	the handle is made of four layers – the first is nylon, followed by two layers of different coloured polypropylene and finally a thermoplastic elastomer (TPE) grip
Manufacturer	Stanley Tools
Country	UK
Date	1998

This screwdriver consists of four layers of plastic moulded over the metal shank: the first, blue moulding can be seen at the end of the handle; the shiny black area is the second layer; the yellow graphics are the third; finally, the black grip.

referred to as robot transfer, involves a component being produced first and only afterwards being transferred to another mould for a second material to be added.

There are also other forms of insert moulding that, instead of injection moulding, use compression (p.174), contact (p.152) and rotational (p.137) moulding.

Volumes of production
High-volume production process, typically above 100,000 units.

Unit price vs capital investment
An economical process when compared with manual assembly of the different materials.

Speed
Depends on the product. Thin-walled products cool very quickly, but the type of plastic and overall component design are also important factors.

Surface
Depends on the moulding process used, but is comparable to injection moulding (see p.196) but, in with insert moulding, surface materials may be introduced that can enhance the finish, for example the extra grip on a toothbrush handle.

Types/complexity of shape
Since this type of insert moulding is based on injection moulding, the same possibilities and restrictions apply, although the shape of the insert itself will partly dictate the shapes achievable.

Scale
It is possible to achieve products of greatly varying size depending on the type of injection moulding used.

Tolerances
Can be very high, because injection moulding can achieve tolerances of ±0.1 millimetre.

Relevant materials
Any combination of materials, including thermoplastic and thermoset polymers. Depending on the combination of materials, different layers may bond chemically to varying degrees. However, thermosets and thermoplastic elastomers (TPEs), for example, generally do not provide a chemical bond.

Typical products
One of the key features of combining different materials is that you are able to bring together multiple functions in a single component. For instance, it is possible to have movable joints and decorative features over a flexible yet strong core, without the additional assembly costs. Typical products that are made through this process include toothbrushes, screwdrivers, razors and housings (of, for example, power tools with rubber grips).

Similar methods
In-mould decoration (p.212).

Sustainability issues
This process can eliminate the need for several stages of production and subsequent energy use by reducing everything into one single step in one location, making the most efficient use of transportation and energy. However, recycling becomes more difficult when two or more materials are combined, as they need to be separated before they can be reprocessed.

Further information
www.engel.info
www.bpf.co.uk
www.mckechnie.co.uk

Robot Transfer Method

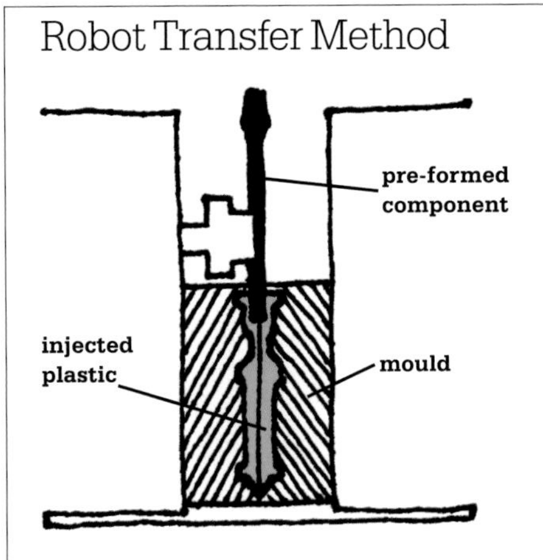

1 Plastic is injected over a preformed component, in this case the metal shaft of a screwdriver.

2 The moulded plastic part (together with the shaft) is removed by robotic arms and transferred into a separate die.

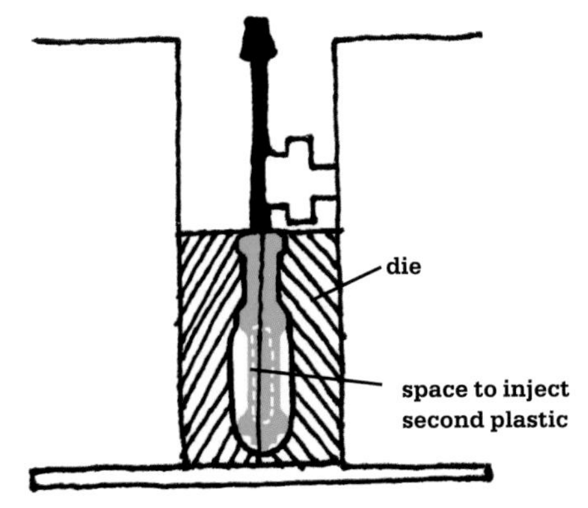

3 At this point, a second plastic is injected over the original moulding. This process can be repeated as many times as necessary to build up the required number of materials.

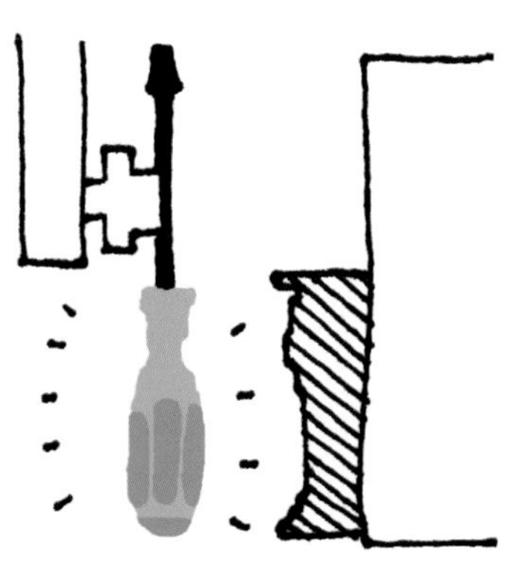

4 The finished component is removed from the mould.

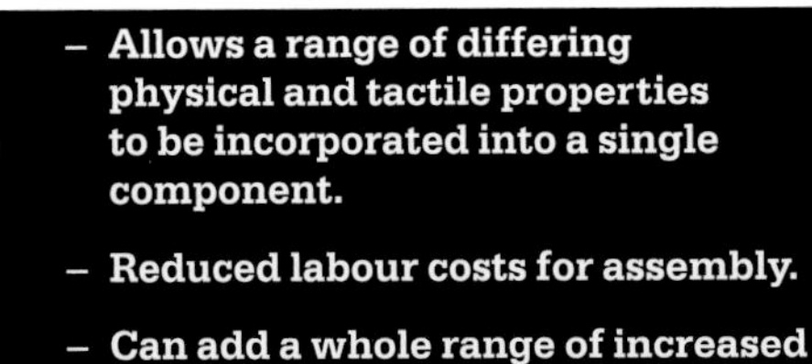

- Allows a range of differing physical and tactile properties to be incorporated into a single component.
- Reduced labour costs for assembly.
- Can add a whole range of increased functionalities.

- High tooling costs.
- Requires an advanced degree of knowledge on how to combine the various materials, and on subsequent design considerations such as shrinkage and the stresses of one material over another.

Multi-Shot Injection Moulding

In contrast to traditional injection moulding where each component is produced separately as a single piece, multi-shot injection moulding makes it possible to mould several pieces and assemble them in a single cycle, resulting in a finished product in just one run. The process even allows for products with moving parts such as caps, handles and hinges to be created within the same moulding cycle. Generally, the manufacture of a product with several components can be a lengthy process that requires each piece to be moulded separately, often in different locations, along with an additional manual assembly to complete production. Multi-shot injection moulding makes production

Product	**Garden secateurs**
Materials	Thermoplastic Elastomer (TPE) grip, polypropylene handle, steel blades
Manufacturer	Fiskars
Country	Finland

The cut-away of these secateurs show how the grey TPE is injection moulded within the black PP handle.

a lot less messy by eliminating all these secondary processes and incorporating them into one, which in turn significantly reduces lead time as well as costs.

As with anything simple, a great deal of set-up is required in order to maximise the process. Every aspect must be considered, from how the combination of materials will react with one another to the type of machine that should be used. The design of the moulding process is very creative because of the flexibility of the machine processing. The designers examine several different ways of making a part as there is no single solution. Some methods may offer cost savings while others may allow for flexibility in designing parts.

A multi-shot injection moulding machine is automated and consists of robotic arms that are controlled by a computer and move the component from one mould cavity to another.

+

- Reduced production time.
- Reduced unit costs.
- Multi-component part produced in single cycles.
- Up to four different materials can be moulded in one cycle.
- A variety of functional/ decorative features, including graphics, text and grips, can be applied during the process

–

- Needs a great deal of production planning.
- It is likely that the original design will require some alterations before it is suitable for manufacture.
- The robot sometimes has difficulty picking up and placing complex components, which can result in mouldings being dropped and misplaced.

Volumes of production
There are small injection moulders that produce simple components of 5000 units or less. However, the minimum quantity is generally accepted to be 10,000 units.

Unit price vs capital investment
The investment cost for machinery and design planning is high. However, the overall cost is lower in comparison with traditional techniques due to a reduction in storage, assembly costs, transportation of components and use of materials.

Surface
The steel mould allows for surfaces that range from spark eroded to highly glossy. The location of ejector pins in the mould needs to be considered when designing a part as they will leave a small indented circle. Parting lines where the various parts of the mould come together also need to be considered.

Types/complexity of shape
If production volumes are particularly high multi-shot injection moulding can be used to form highly complex parts. However, elements like undercuts, variable wall thicknesses, inserts and threads significantly add to the cost of tooling.

Scale
Micro injection moulding is a specialist area and certain manufacturers specialise in parts that are often less than 1 millimetre. For large-scale products like garden chairs it is worth considering gas-assisted injection moulding or, for large wall thicknesses, reaction injection moulding (RIM).

Tolerances
± 0.1 millimetre.

Relevant materials
The process is suited to the thermoplastics family, but consideration needs to be given to the compatibility of various types when mixing them in a component.

Typical products
Medical and healthcare; automotive; telecommunications; electronics; appliances; cosmetics.

Similar methods
For other methods that combine materials see over-mould decoration (p.214), insert moulding (p.206), in-mould decoration (p.212).

Sustainability issues
As all parts are produced in one process, energy use is very efficient and transportation between locations for additional manufacturing is eliminated. However, it is very difficult to separate materials in multi-material components, which makes them problematic to recycle. Therefore, consideration should be given to reducing the number of material types.

Further information
www.mgstech.com
www.fiskars.com

In-Mould Decoration

As the name implies, this is not a method of production as such, but, rather, in-mould decoration was developed as an economical way of adding decorative surfaces to injection-moulded plastic parts. It offers a way of eliminating the necessity of having to print directly onto a part in a separate, post-forming process. This manufacturing technique is becoming more and more important with the growing market for electronic products, which makes increasing use of graphics for keypads, product branding and the personalising of portable consumer products.

The process begins with the printing of the graphic onto a polycarbonate or polyester film, known as a 'foil'. Depending on the shape of the component to be moulded, the foil is fed on a ribbon into the mould (or is cut and individually inserted, if the part is curved). The process is also suitable for shapes with compound curves, but in this case the foil needs to be moulded to shape before being inserted into the mould.

One of the uses for in-mould decoration is as an alternative to spraying or moulding parts in specific colours. It is a way of applying colour to products to ensure consistency between mouldings in different materials, where exact colour matches are hard to achieve. An example of the use of this process can be found in the back and front mouldings of a mobile phone, where the back is moulded in one material and the front in another.

Product	Demonstration sample masks
Materials	The process has optimum effect on PBT plastic
Manufacturer	IDT Systems
Country	UK

Shown left are demonstration masks produced by IDT Systems to illustrate the depth of penetration that is possible with this system. These particular samples are made from PBT plastic with a translucent film.

Volumes of production
Suited to mass-production.

Unit price vs capital investment
In-mould decoration is very cost-effective compared with painting or spraying parts in a separate process.

Speed
As you might imagine, inserting the film has a slightly negative effect on the overall cycle times, but this step can be automated and it needs to be considered in relation to the time it would otherwise take to decorate a part by, for example, painting it.

Surface
Different films can be used to give a variety of finishes ranging from the functional to the decorative.

Types/complexity of shape
In-mould decoration can be used on both simple and complex compound curves.

Scale
As injection moulding (see p.196). It is possible to make very small parts, but the shapes need to be very simple.

Tolerances
Not applicable.

Relevant materials
Polycarbonate, acrylonitrile butadiene styrene (ABS), polymethyl methacrylate (PMMA), polystyrene and polypropylene.

Typical products
In-mould decoration is not just limited to text-based graphics, but is also used to produce colour on mouldings and to add surface patterns. One of the most interesting (albeit invisible) foils that can be applied is a form of 'self-healing' skin. This protective layer helps to keep handheld products, such as mobile-phone casings, shiny and free from scratches. Other applications include decorative mobile-phone covers, machine fascia, digital watches, keypads and automotive trim, to mention just a small selection of products.

Similar methods
Over-mould decoration (p.214) is similar, but involves the application of materials, rather than foils, to a moulding. Sublimation coating is another alternative, although it is applied as a secondary process after moulding and is particularly suited to engineering polymers such as nylon.

Sustainability issues
By combining decoration with production, energy consumption can be significantly lowered as there is no additional machining, power use and transportation. In addition, the decorative films can be used to enhance and protect the material surface, which can prolong the life of the product.

Further information
www.autotype.com.sg
www.filminsertmoulding.com
www.idt-systems.com

+

- **Cost-effective to customise parts and provide customer differentiation without re-tooling.**
- **Allows for virtually any colour, image and even surface texture to be added as a skin.**
- **Equally suited to short and long production runs.**
- **Films can be used to offer surfaces that are scratch-, chemical- and abrasion-resistant.**

–

- **Incurs additional moulding costs because of the complexity of the mould required to accommodate a film or foil.**

Over-Mould Decoration

Over-mould decoration is not really a method of production in is own right, but rather an extension of standard injection moulding (see p.196), as part of a two-step process. What is particularly noteworthy about it is that it can give plastic components an almost craftlike quality by the way it cleverly allows a different material to cover the plastic in the mould.

If you were to see a mobile-phone casing, for example, that has a small patch of fabric wrapped onto its surface, you might find it an interesting combination and imagine that the addition of the fabric would require a whole new process involving someone fixing the fabric by hand onto the plastic moulding. In reality, this would be labour-intensive and expensive. Inclosia Solutions, a branch of Dow Chemical, has come up with the technology to combine plastic with a range of other materials during the moulding process itself, eliminating the need for any secondary finishing.

The benefit of this type of manufacturing is that it provides designers with a new set of materials, surfaces and finishes to challenge traditional notions of plastic-moulded products. Instead of electronic products having the same all-over plastic skin, they can have warm, tactile surfaces that are closer to textiles or to crafted materials such as wood. The process offers the possibility of extending our perception of products beyond the current boundaries of identical, mass-produced plastic, making it feasible for products to be 'dressed' and become more integrated with our clothes, furniture and jewellery.

Product	**E-Go laptop**
Designer	Marcel van Galen
Materials	fabric over a plastic moulding
Manufacturer	Tulip
Country	The Netherlands
Date	2005

The internal plastic moulding of this laptop computer can be over-moulded with a range of different materials to suit varying consumer markets. The leather and fabric 'skins' allow for consumer electronics to be marketed in the same way as more fashion-led products, such as handbags.

Volumes of production
High-volume production process.

Unit price vs capital investment
Unit price for components is more than that for standard injection moulding (see p.196). Tooling costs are higher due to the need to incorporate a second material.

Speed
Because it is a two-step process that involves forming a material over a pre-formed component, it is slower than some similar methods of multi-component (that is, two-shot) moulding.

Surface
The key feature here is that it allows for secondary 'skins' to be applied over plastic mouldings, so the surface is determined by the material you choose as the covering.

Types/complexity of shape
Because of the secondary material, over-mould decoration is best suited to flat surfaces and those with a deep draw.

Scale
The largest standard size is approximately 300 by 300 millimetres.

Tolerances
Depend on the shrinkage of the various materials.

Relevant materials
A variety of thin materials can be used to over-mould, such as aluminium sheet, leather, fabric and thin wooden veneers.

Typical products
Over-moulding has been used in a range of products that fall into the category of personal mobile technology, including mobile phones, Personal Digital Assistants (PDAs) and laptop computers.

Similar methods
In-mould decoration (p.212) and insert moulding (p.206).

Sustainability issues
Over-mould decoration requires an additional stage of processing to apply decoration to a moulded product which, of course, increases energy consumption. However, the enhanced decorative surface could help to extend the product's lifespan through increased value perception.
As with any multi-material component, recycling can be problematic because of the need to separate materials.

Further information
www.dow.com/inclosia
www.filminsertmoulding.com

+

- **Automated method of covering plastic-moulded components with a second soft, or decorative, material.**
- **Cost-effective alternative to hand assembly.**
- **Compatible with most engineering thermoplastics and elastomers.**

–

- **Although over-moulding provides benefits when it comes to surface decoration, it is a two-step process, adding to unit price.**
- **Can require trial and error when using untested materials.**

Metal Injection Moulding (MIM)

A variation on standard injection moulding (which uses plastics; see p.196), metal injection moulding (MIM) is a relatively new way of producing complex shapes in large numbers from metals that have a high melting point, such as tool steel and stainless steel that would not be suitable for high-pressure die-casting (see p.219). The process is limited by the suitability of the metal powders that are used as the raw material, which need to be particularly fine.

MIM involves more processes than plastic injection moulding, because of the necessity of adding binders to the metal. The various companies involved in using this technology each have their own unique binder systems but typically

Product	**engineering components [pen nib for scale only]**
Materials	low-alloy steel and stainless steel
Manufacturer	Metal Injection Mouldings Ltd, part of PI Castings
Country	UK
Date	first produced in the UK in 1989

This range of small-scale, complex engineering components is typical of the type of products that are made using MIM. It offers us the chance to create precise, solid metal products from metals with high melting points that cannot easily be formed by casting. The strength and hardness of these components offers several advantages over other forms of metal production.

the binders, which can account for 50 per cent of the compound, are made from a variety of materials, including wax and a range of plastics. They are mixed in with the metal powders to produce the moulding compound.

Once the shapes have been moulded, the binder is no longer needed and is removed from the metal particles. What is left is then sintered (see p.168), which shrinks the component by about 20 per cent.

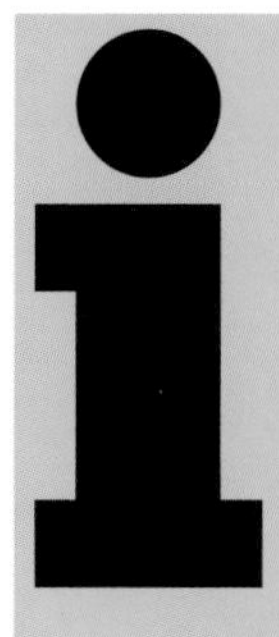

Volumes of production
In order to justify the set-up and tooling costs, high-volume production is needed – a minimum of 10,000 units.

Unit price vs capital investment
High capital investment but a very low unit price.

Speed
The actual injection of material is similar to that of standard injection moulding in plastics (see p.196), but the sintering and the removal of the binder add time and expense to the process.

Surface
The process gives an excellent surface finish on components and has the ability to produce fine detail.

Types/complexity of shape
Highly complex shapes similar to those obtainable through standard plastic injection moulding. These can also be enhanced by the use of multi-cavity tooling.

Scale
MIM is currently capable of producing only small parts for use in larger products.

Tolerances
The MIM process can achieve a general tolerance of ±0.10 millimetres.

Relevant materials
MIM is economical for producing large numbers of complex components with a range of surface finishes. It can be applied to a range of metals: bronze, stainless steel, low-alloy steels, tool steels, magnetic alloys and alloys of low thermal expansion.

Typical products
Surgical and dental tools, computer components, automotive parts, casings for electronics and consumer products (mobile phones, laptops, PDAs).

Similar methods
Although die-casting in metal (p.219) is possibly the closest to MIM in terms of production quantities and complexity of shape achievable, the key difference between the processes is in the ability of MIM to work with metals with high melting points, such as low-alloy steels and stainless steel.

Sustainability issues
The additional processing and heating cycles significantly increase energy consumption in comparison with traditional plastic injection moulding. Compared to casting or metal machining, there is practically no excess or scrap material, which helps to reduce waste and energy use from secondary processing. The high-temperature nature of the materials means they are less likely to feed into the recycling stream.

Further information
www.mimparts.com
www.pi-castings.co.uk
www.mpif.org

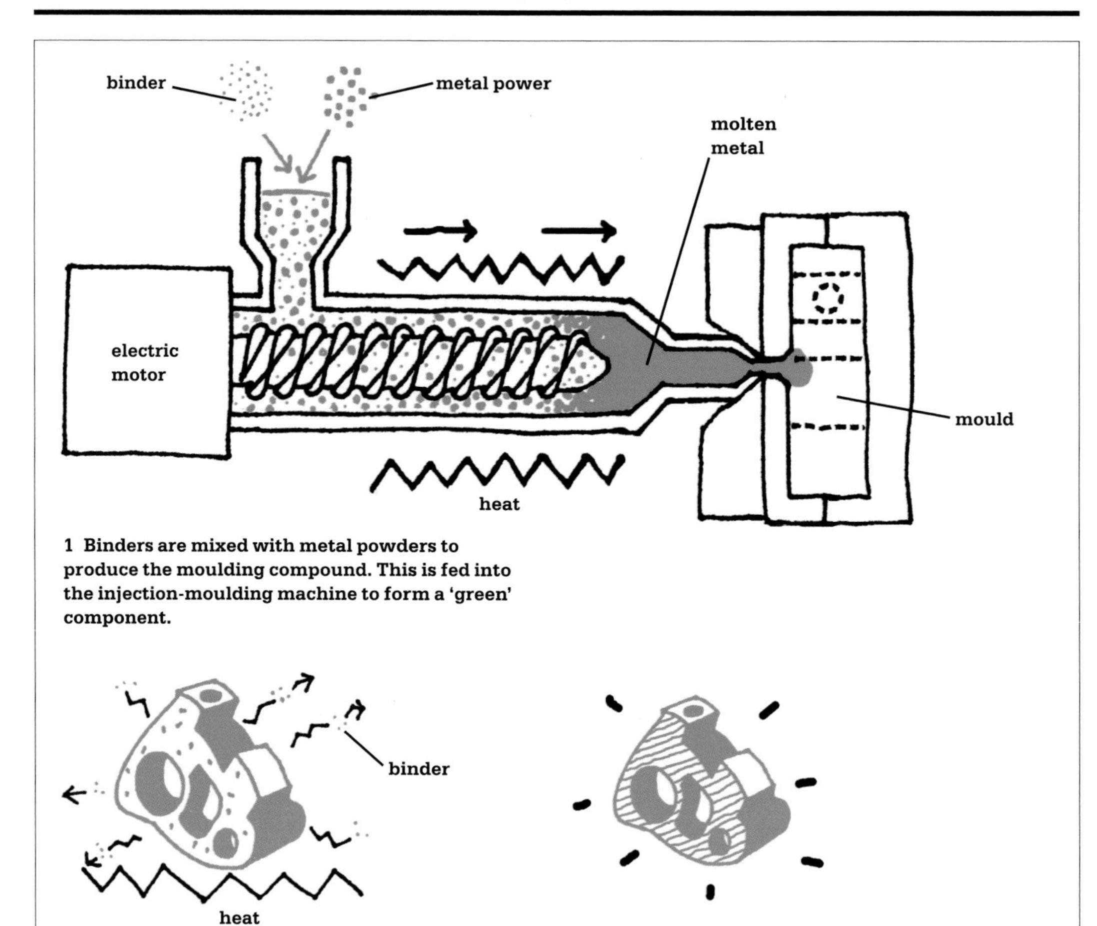

1 Binders are mixed with metal powders to produce the moulding compound. This is fed into the injection-moulding machine to form a 'green' component.

2 After the shape has been moulded, the binder is removed from the metal particles and discarded. This step is achieved in a number of ways depending on the specific manufacturer.

3 What is left is sintered to weld the metal particles together. This shrinks the component by about 20 per cent.

+

- Can be used to form high-temperature alloys.
- Used to form complex shapes.
- Cost-effective for large numbers.
- No post finishing required.
- Parts have exhibited good strength.

–

- Low overall part size.
- Compared with standard injection moulding in plastic (see p.196), only a limited number of manufacturers can offer MIM.

High-Pressure Die-Casting

High-pressure die-casting is one of the most economical methods of producing metal components with complex shapes. It is the process to use if you want to produce high volumes of intricate components. In this sense, it is similar to metal injection moulding (MIM) (see p.216), but its main advantage over MIM is that it is suitable

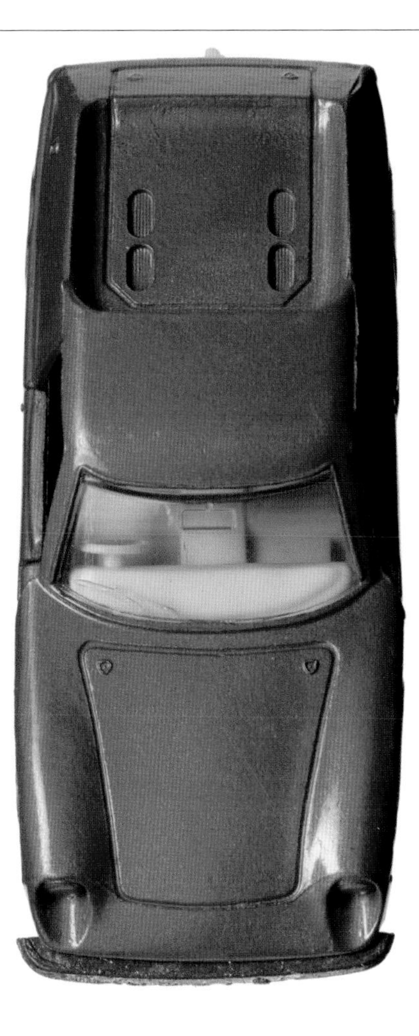

Product	**Matchbox Lotus Europa**
Materials	zinc
Manufacturer	Matchbox
Country	UK
Date	1969

Die-cast metal toys are part of many people's childhood memories. The ability of die-casting to create fine, complex details is well illustrated by the clearly legible text on the underside of my son's toy car.

for metals with low melting-points where no sintering is required.

The process involves molten metal being poured into a reservoir, where a plunger forces the liquid, under high pressure, into a die cavity. The pressure is maintained until the metal solidifies, at which point small ejector pins push the components out of the die. Just as in injection moulding (see p.196), die-casting moulding dies are made in two halves.

Volumes of production
High-pressure die-casting is strictly for high-volume production.

Unit price vs capital investment
Economical unit price is obtained by mass-producing highly complex parts, which eventually drives down the relative cost of the expensive tooling that has to be designed to withstand repeated injection of molten metal at high pressure.

Speed
Fast, although the removal of flashing as a separate process adds to the time.

Surface
Excellent.

Types/complexity of shape
Ideal for producing complex, open-walled parts in metal, especially those with thin wall sections. Unlike investment casting (see p.224), high-pressure die-casting requires draft angles.

Scale
Up to a maximum weight of approximately 45 kilograms for an aluminium component.

Tolerances
Reasonably high level of tolerance, but shrinkage can sometimes be problematic.

Relevant materials
Metals with a low melting temperature, such as aluminium and zinc, which are by far the most commonly used materials. Others include brass and magnesium alloys.

Typical products
Chassis for a range of electrical products such as PCs, cameras, DVD players, furniture components and wet-shaver handles.

Similar methods
Investment casting (p.224) and sand casting (p.228), which allow the casting of larger parts and require les capital investment, but demand higher tolerances. Gravity die-casting is a much older process and is employed on a much smaller scale of production than high-pressure die-casting.

Sustainability issues
The low melting points of the metals used in high-pressure die-casting require lower temperatures and quicker cycles than some other processes, such as MIM, so the process consumes less energy and releases fewer emissions. However, excess flash material after casting requires additional trimming, which adds to energy use and waste. At the end of their use, this can be reclaimed and recycled to reduce the use of virgin metals.

Further information
www.castmetalsfederation.com
www.diecasting.org

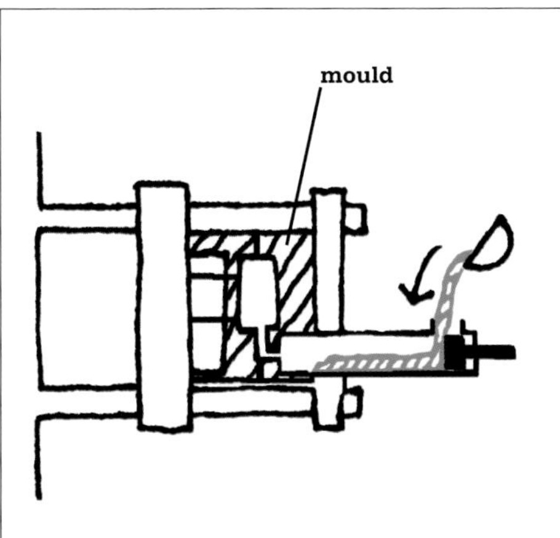

1 Molten metal is poured into a reservoir.

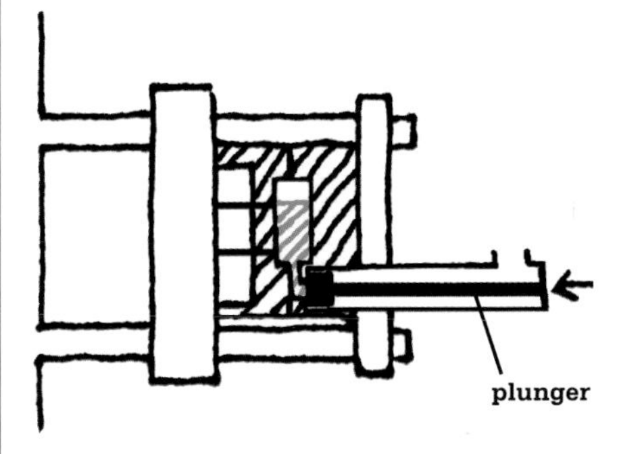

2 A plunger forces the liquid under high pressure into a die cavity.

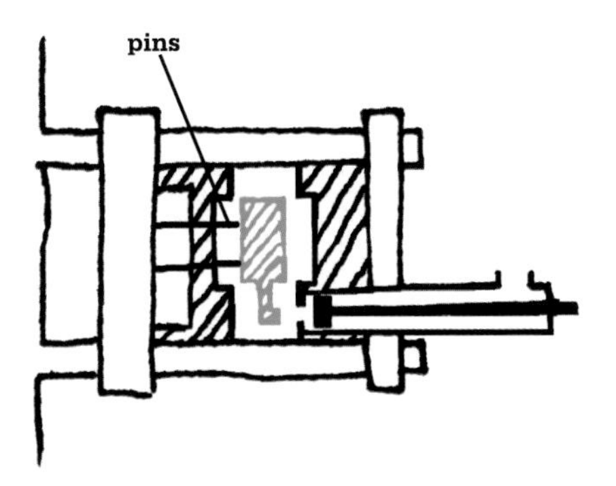

3 The pressure is maintained until the metal solidifies, at which time small ejector pins push the components out of the die.

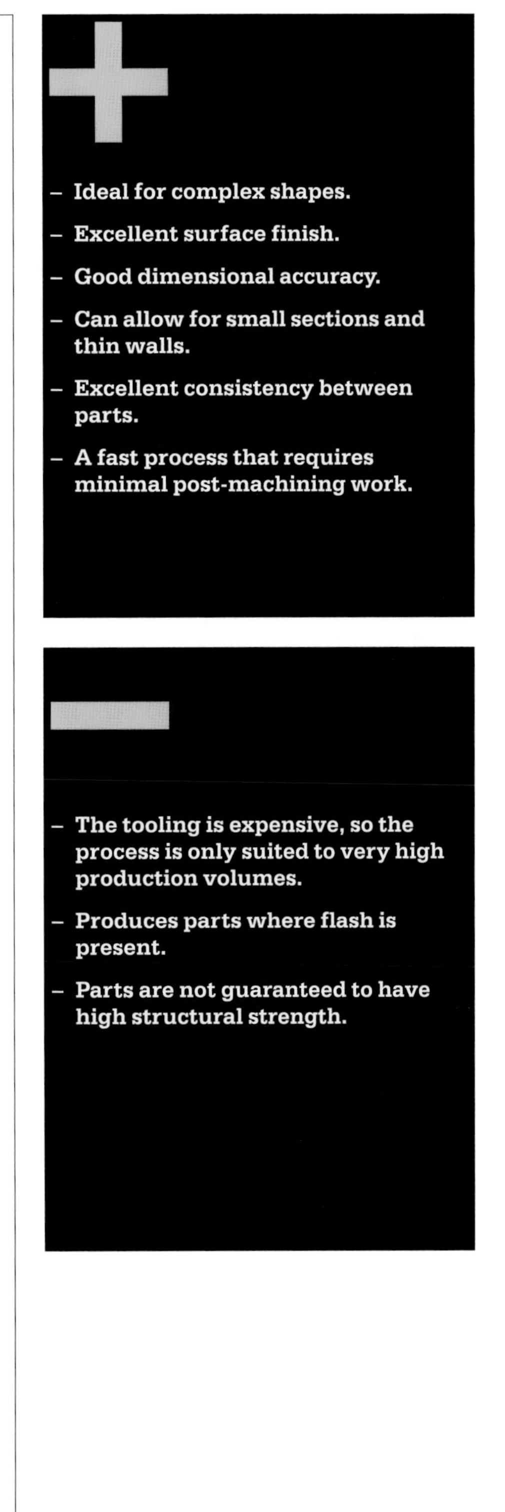

Ceramic Injection Moulding (CIM)

Ceramics provide a hard, wear- and corrosion-resistant solution for many engineering challenges that other materials are rarely able to meet. The injection moulding of ceramics allows for complex shapes to be created in a process that borrows a great deal from the injection moulding of plastics. The technique is suitable for scales of production from one-off research prototypes to mass-produced components for commercial products. Ceramic injection moulding (CIM) has been especially effective within the medical field, to make components for pacemakers and surgical instruments that need to be micro-miniature, with exceptional tolerances, as well as being biocompatible.

The most suitable type of ceramic powder is chosen and mixed with a binder that allows the mixture to flow and become mouldable. The binder is a key aspect of the moulding process and is used because it has a lower melting point than the ceramic powder, which enables the two materials to be separated at a later stage. A specially made machine with greater corrosion strength than traditional injection moulding machines is used to feed the ceramic and binder mix into the mould cavity. The corrosion strength is required because of the abrasiveness of the ceramic that is being moulded.

Once the component has cooled the mould is heated until it is just high enough to melt the binder material (but not the ceramic) causing it to evaporate and leave behind just the ceramic material. The finished part can be sintered or undergo hot isostatic pressing (HIP) in order to remove any stresses caused during moulding and provide further strengthening.

Product	**Dental bracket used in braces [pencil sharpener is for scale]**
Materials	99 per cent Alumina
Manufacturer	Small Precision Tools
Country	US
Date	2010

The image illustrates a specific technology from Small Precision Tools that allows for moulding ceramics such as dental brackets for braces at a micro scale.

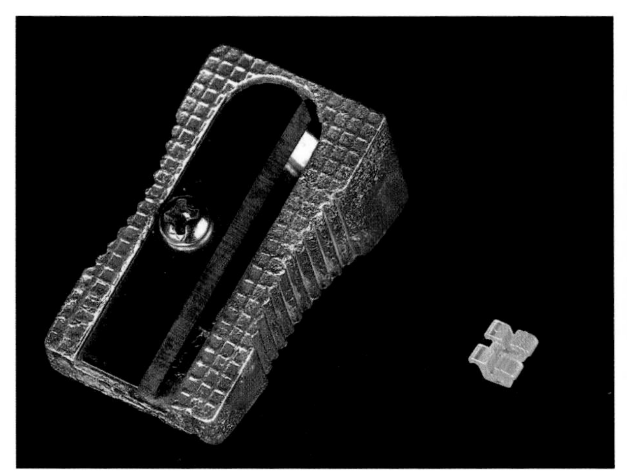

Volumes of production
The process allows for complex ceramic parts to be produced in tens of thousands.

Unit price vs capital investment
Producing moulds means the initial set-up cost is very high. Unit price is reduced when the scale of production is increased.

Speed
Multiple components are produced in the same production run to optimise the timescale: it can take many days for a part to go through the various stages.

Surface
The nature of ceramic materials results in a fine, stone matt surface.

Types/complexity of shape
The types of shape have similar restrictions to those of injection moulded plastics, where the main considerations should be undercuts and the removal of components from the mould.

Scale
The component featured here from Small Precision Tools illustrates the tiny scale that this process is capable of. Components produced with this process are typically measured in 1 or 2 millimetres, but components can be produced that would fit through the eye of a needle.

Tolerances
These vary depending on the type of material but ± 0.005 millimetres can be obtained.

Relevant materials
A range of ceramics, including zirconium oxide, silicon carbide and aluminium oxide.

Typical products
CIM is particularly suited to small engineering components as ceramics have exceptional resistance to wear and corrosion, and chemicals, and are biologically inert. This makes the process suitable for a range of parts including dental implants.

Similar methods
Injection moulded plastics (p.196)

Sustainability
The main consideration in this multi-stage process is the use of heat during sintering and the removal of the binder.

Further information
www.smallprecisiontools.com

- **Complex and intricate parts that would otherwise be difficult or even impossible to produce in ceramics can be achieved.**

- **Production of moulds can be costly and time consuming.**

Investment Casting
AKA Lost-Wax Casting

The name 'investment casting' is taken from the idea that the process involves 'investment' in a sacrificial material, and it is characterised by its ability to produce highly complex shapes. The process has been around for thousands of years, with evidence of its use by the ancient Egyptians. In essence, it involves a wax shape being dipped into a ceramic liquid which itself forms a thick enough skin to hold molten metal once the wax has been melted away. Because the ceramic mould is broken to reveal the finished object, it is possible to get away with all kinds of undercuts and complex shapes that would not be possible to achieve with a rigid mould.

The first stage involves the manufacture of a die (usually made

wax pattern

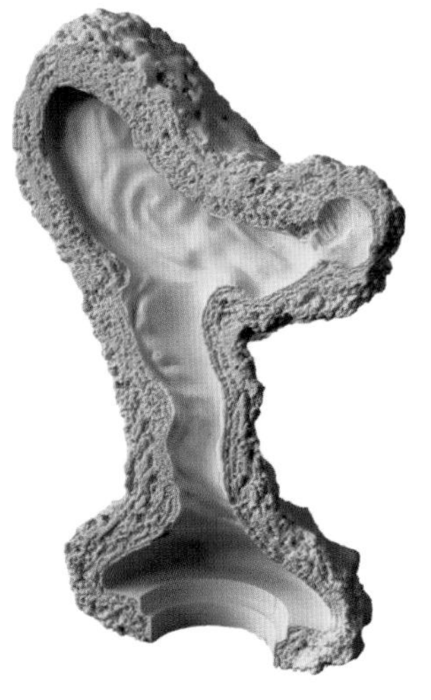

ceramic shell

finished product

Product	**Spirit of Ecstasy car hood ornament**
Designer	Charles Robinson Sykes
Materials	stainless steel
Manufacturer	Polycast Ltd
Country	UK
Date	1911

These images illustrate three of the stages of investment casting for this highly recognisable figure. They also offer an excellent example of which method of production to choose when the fashion for ornate, decorative figurines returns to contemporary design.

from aluminium, but a polymer can also be appropriate), which is repeatedly used to obtain the wax replica patterns. Multiple patterns are produced that are assembled onto a wax runner, which forms a structure that resembles a kind of tree. This assembled runner is then dipped into ceramic slurry, which is dried to form the hard ceramic skin. The dipping is repeated until sufficient layers have been built up. The runner is then

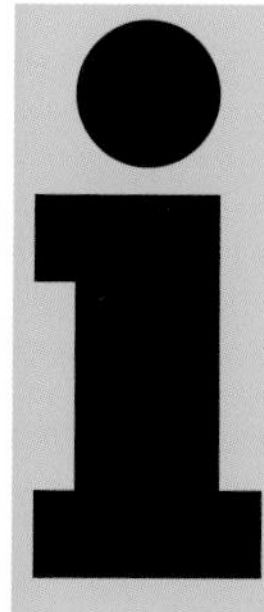

Volumes of production
Depending on size, it is possible to have several hundred small parts on a tree, which can be cast in one pour. Larger items are made with only one per tree. Investment casting is a process that allows for small runs of below a hundred, as well as runs of up to tens of thousands.

Unit price vs capital investment
The tooling is far cheaper than that needed for high-pressure die-casting (see p.219), which means lower capital investment. Depending on the size of the final component, multiple castings can be produced on the same tree to increase cost-effectiveness.

Speed
Slow, requiring a number of steps to be completed for each component.

Surface
Good surface finish, but this is largely dependent on the surface of the pattern.

Types/complexity of shape
Unlike in high-pressure die-casting, which requires draft angles, components made by investment casting can be highly complex. This is the main advantage the process has over other methods of forming.

Scale
Anything from 5 millimetres to about 500 millimetres long, or up to 100 kilograms.

Tolerances
High.

Relevant materials
A wide variety of ferrous and non-ferrous metals.

Typical products
Anything from sculptures and statues to gas turbines, marine shackles, jewellery and medical tools. One example with an extremely high profile is the 'Spirit of Ecstasy' that sits on the bonnets of Rolls-Royce cars.

Similar methods
High-pressure die-casting (p.219), sand casting (p.228) and centrifugal casting (p.161).

Sustainability issues
After being smashed, the sacrificial ceramic can be collected and heated back into its slurry state to prevent waste and to reduce further use of raw materials. The process can be relatively energy intensive as obtaining the finished product involves stages of heating and processing. Some foundries still use alcohol-based binders in the shell, which may pose a threat to the environment when disposed of. The main issue with any metal-casting technique is the heat used during the process.

Further information
www.polycast.co.uk
www.castmetalfederation.com
www.castingstechnology.com
www.pi-castings.co.uk
www.tms.org
www.maybrey.co.uk

placed in an oven to melt the wax so that it can be poured out before the ceramic is fired. The ceramic shells are now strong enough to allow molten metal to be poured into them. After cooling, the ceramic can be broken away and each part may be removed from the tree inside.

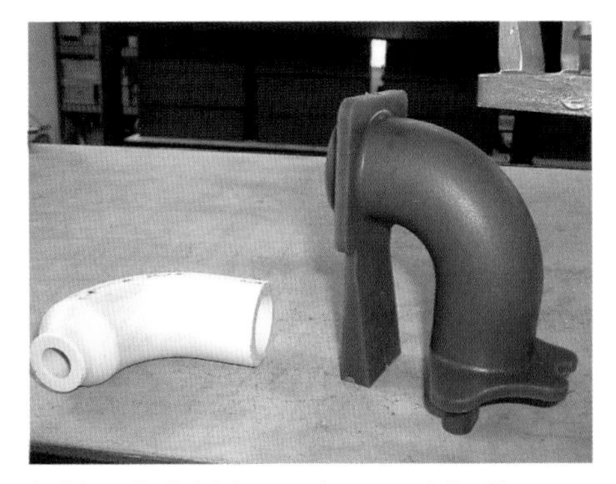

1 After the initial manufacture of the die, a wax pattern (on the right) is produced.

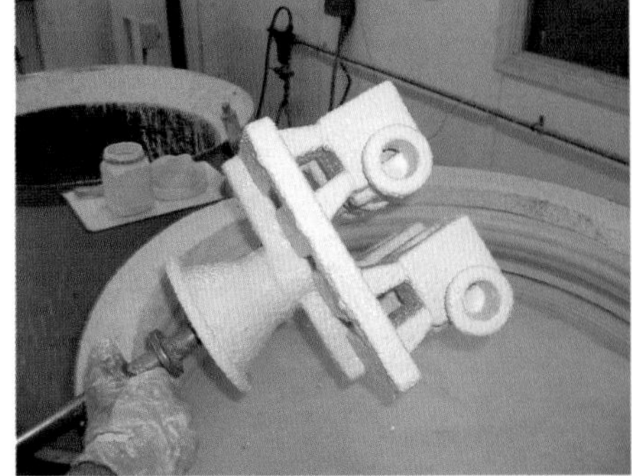

2 This image, featuring a different pattern, shows a simple set of four components being dipped into the ceramic slurry.

3 A typical set-up showing a number of wax components on a simple runner before being dipped into the slurry.

4 The ceramic shell filled with metal (with a finished component held next to it for comparison).

5 This is the stage where the dried ceramic is removed and discarded revealing the final component.

6 A final component shown with the original wax pattern.

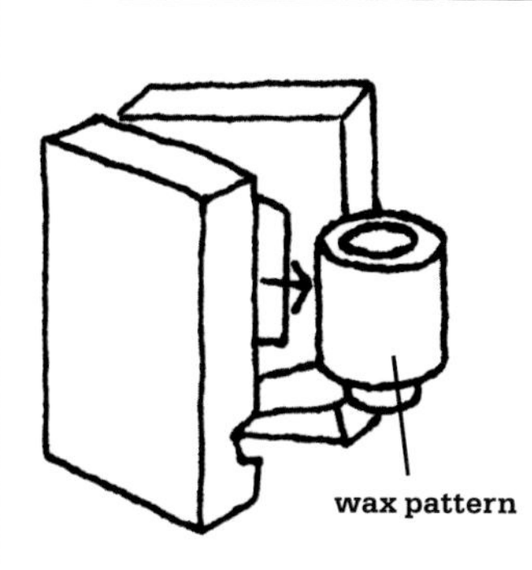

1 Wax patterns are made using an aluminium die, which is reused to obtain the required number.

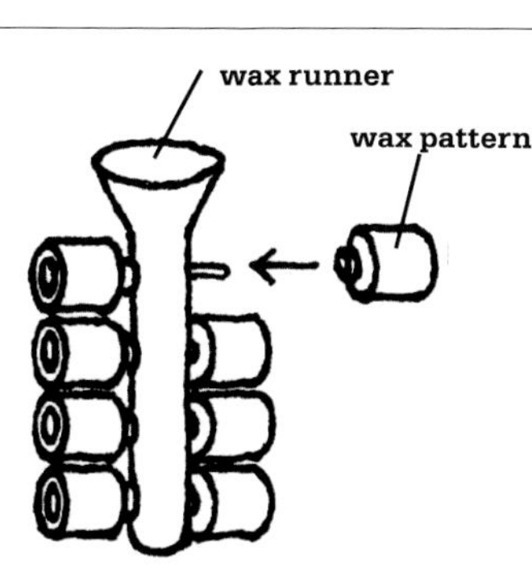

2 The individual wax patterns are assembled onto a wax runner.

3 The assembled runner is dipped into ceramic slurry and dried to form the hard ceramic skin. The process is repeated until sufficient layers have built up.

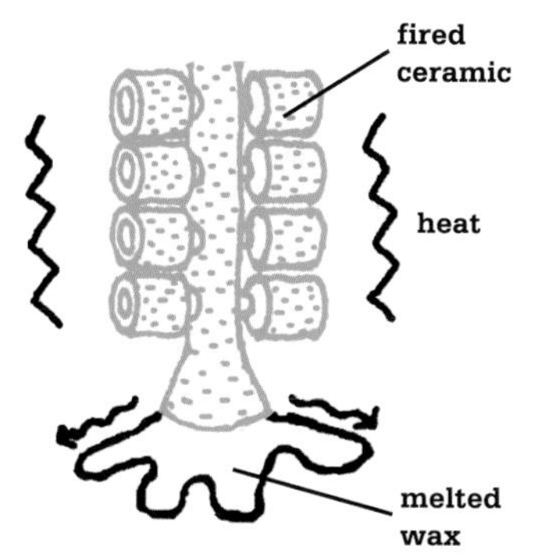

4 The runner is placed in an oven to melt the wax so that it can be poured out before the ceramic is fired.

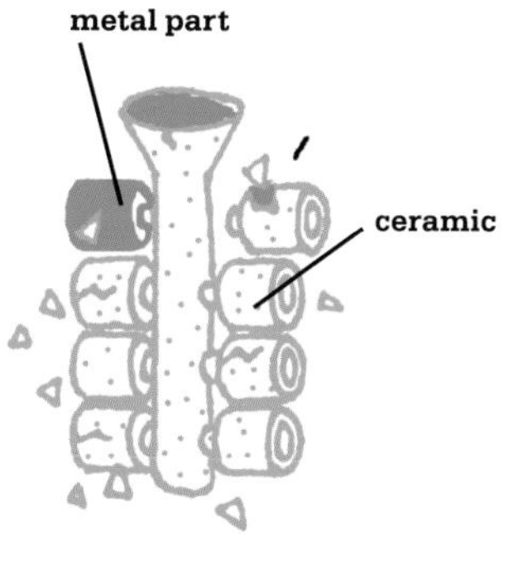

5 Molten metal is poured into the fired ceramic shells. After cooling, the ceramic is broken away and each metal part can be removed from the tree inside.

6 The finished casting.

+

- Complex shapes with hollow cores are possible.
- Weight savings due to the ability to form hollow cores.
- A process for high accuracy.
- Eliminates post-process machining operations.
- Freedom of design.

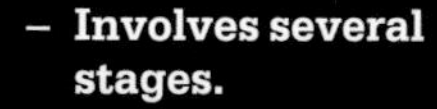

–

- Involves several stages.
- Some foundries still use alcohol-based binders in the shell, which may pose a threat to the environment.

Sand Casting

including CO² silicate and shell casting

Of the many attributes of sand, one that stands out is the fact that it is a refractory material. This means that it can withstand extremely high temperatures, and thus easily accommodate molten metals for casting. There are various forms of sand casting, the differences lying mainly in the quantity of the components that is required, but all rely on the very simple principle of making a pattern (or duplicate) of the finished part. This duplicate is embedded in a compacted mixture of sand and clay and then removed, leaving a cavity into which a molten metal can be poured. Runners and risers are used in the sand to contain a reservoir of excess molten material. These are, essentially, holes in the sand: the runner allows metal to be poured in; the riser holds any excess molten metal. This is a necessary precaution, because as molten metal solidifies, it shrinks, and at this stage excess metal is drawn into the mould to prevent voids in the casting.

There are several derivatives of this basic principle. These include the use of patterns made from sacrificial materials such as polystyrene foam, which evaporate when the metal is poured in. Wooden patterns are used for small-batch work in foundries, while the process can also be automated in a procedure that uses aluminium patterns and a programmed compaction method.

Other methods include CO² silicate casting and shell casting. The CO² process is a recent development, and it involves the sand being bonded with sodium silicate instead of clay. This is converted into CO² during casting and it can provide greater accuracy because sodium silicate makes a tougher mould. Shell casting uses fine-grained, very pure sand, coated in a thermosetting resin. This means that the mould can be thin walled (as little as 10 millimetres) but it is very strong. Shell casting offers several advantages over conventional sand casting, such as greater tolerance and a smoother surface.

Product	**High Funk table legs**
Designer	Olof Kolte
Materials	aluminium
Manufacturer	First produced by David Design
Country	Sweden
Date	2001

The concept behind these table legs is that the design is sold without a table-top, so that customers can buy legs to fit under the table-top of their choice.

Volumes of production
Sand casting can be used to make a single component or in large production runs.

Unit price vs capital investment
For manual sand casting, the price is dependent on the cost of making a wooden pattern, with the unit price of the component being relatively cheap. Automated processes are expensive, but will obviously produce lower unit costs.

Speed
Compared with high-pressure die-casting (see p.219), this is a fairly time-consuming process.

Surface
Casting in sand gives a surface that is very textured and that needs to be ground and polished if a fine surface is required. Sand casting using polystyrene leaves no parting lines and thus requires less finishing. Shell casting can also provide a greater surface finish.

Types/complexity of shape
By its nature sand is a fragile material to cast with, which means sand casting is best suited to quite simple shapes. However, the large number of processes that has developed allows for the production of complex shapes with varying wall thicknesses and undercuts.

Scale
Compared with other forms of metal casting, sand casting allows for the casting of very large components, but parts require a minimum of 3 to 5-millimetre wall-thicknesses, and they have a comparatively coarse finish.

Tolerances
As is the case with many other casting techniques, it is important to take shrinkage into account when considering the process. Various metals will have different shrinkage rates, but generally no more than approximately 2.5 per cent. Shell casting provides a higher level of dimensional accuracy.

Relevant materials
As a general rule, metals with low melting points, such as lead, zinc, tin, aluminium, copper alloys, iron and certain steels.

Typical products
Car engine blocks, cylinder heads and turbine manifolds.

Similar methods
Comparable, but more expensive, methods include die-casting (p.219) and investment casting (p.224), but on the whole sand casting is capable of producing more intricate shapes.

Sustainability issues
Virgin sand is used for moulding, and can be reused numerous times within the process. However, heat and abrasion from the molten metal eventually cause damage to the sand, making it unsuitable for continued use, and it becomes waste. Fortunately some of this waste sand can be recycled into non-casting applications, but it is most commonly disposed of in landfills. It is estimated that only about 15 per cent of the several tons of foundry sands generated annually are recycled. As with any metal-casting technique, the main issue is the heat used during the process.

Further information
www.icme.org.uk
www.castingsdev.com
www.castingstechnology.com
www.engineersedge.com

1 The cavity in the bottom mould is clearly visible as the top is lowered.

2 Molten metal is poured into the runners.

3 The part is lifted off with the top mould, ready for finishing.

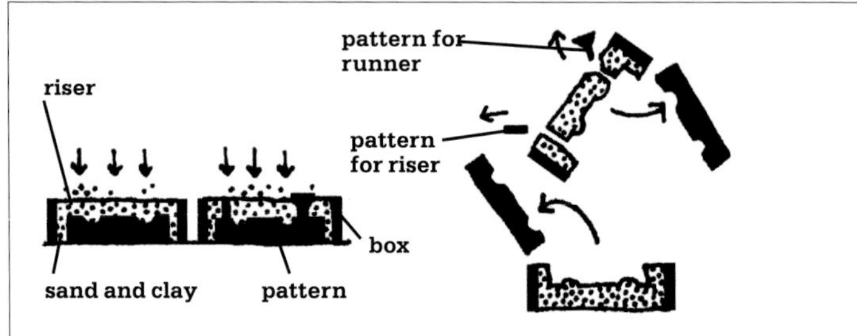

1 First, the original pattern (which includes the runners and risers) is embedded in each of the two halves of the sand box.

2 Once the sand has been compacted, the pattern is removed.

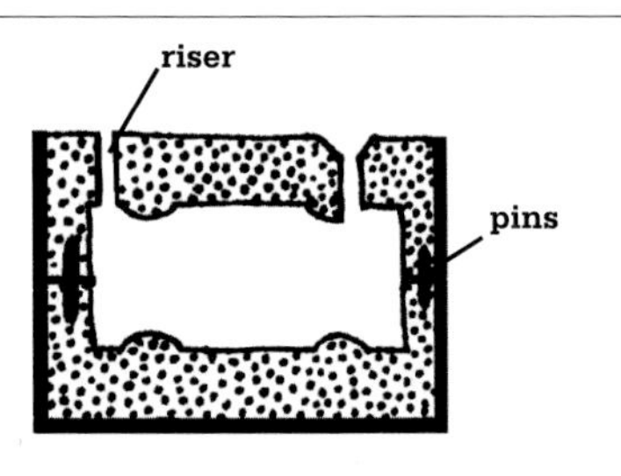

3 The two halves of the sand box are brought together and secured with aligning pins.

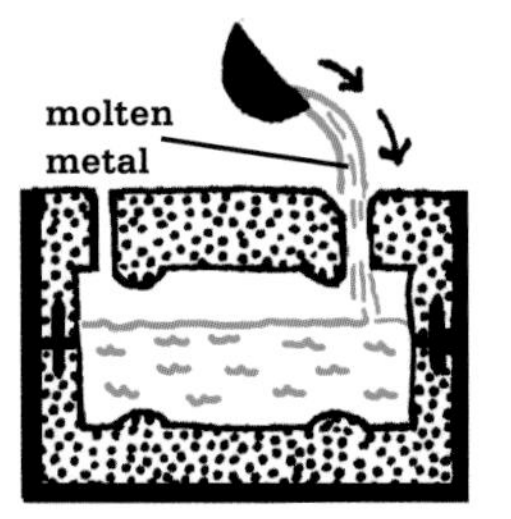

4 Molten metal is poured into the runners, filling the mould cavity.

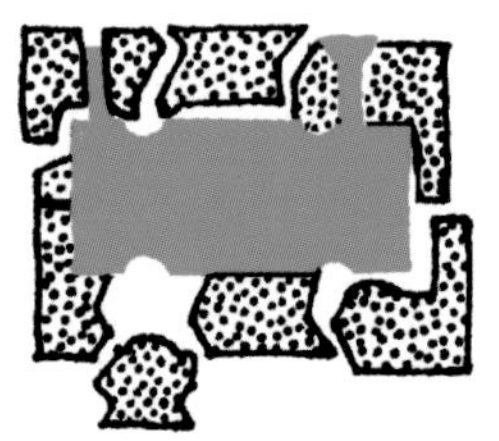

5 Once the casting has cooled, the part is pulled from the sand.

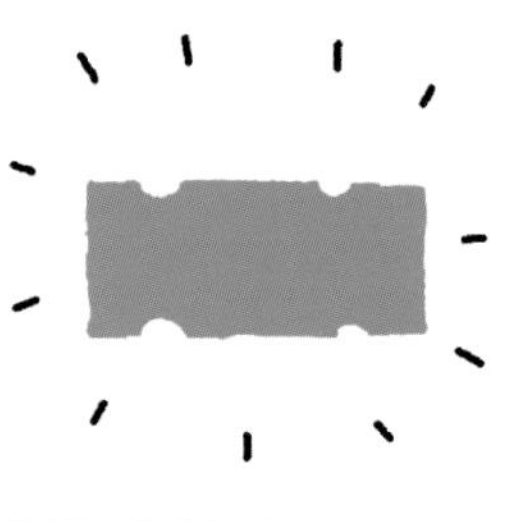

6 The finished part.

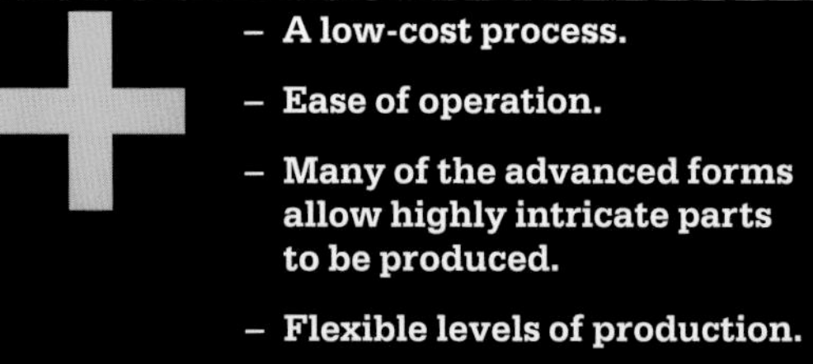

- A low-cost process.
- Ease of operation.
- Many of the advanced forms allow highly intricate parts to be produced.
- Flexible levels of production.

- Can be labour-intensive, with high unit costs when used for small-batch production.
- Parts may require a lot of finishing.

Pressing Glass

Described as the closest thing to 'injection moulding for glass', the pressed glass process makes it possible to mass-produce intricate glass products with detailing on the inside, as well as the outside, of the shapes. This is in marked contrast to glass blowing (see p.116), in which detailing is restricted to the outside surface only. It is possible to trace the staggering boom in the mass-production of all kinds of inexpensive glass products back to the introduction of pressed glass in 1827.

The core of the process involves male and female moulds that are carefully preheated and maintained at a steady temperature to ensure that the hot glass will not stick to the moulds. A gob of gummy, molten glass is squashed between the two moulds, with the amount of space left between the male and female parts determining the thickness of the final component. It is these two moulds – which produce an inner and outer imprint – that allow the shape to be controlled on two surfaces. In large-scale production, the machines typically work on a turntable with a number of stations performing the various stages of production, from filling the mould with glass to the actual pressing.

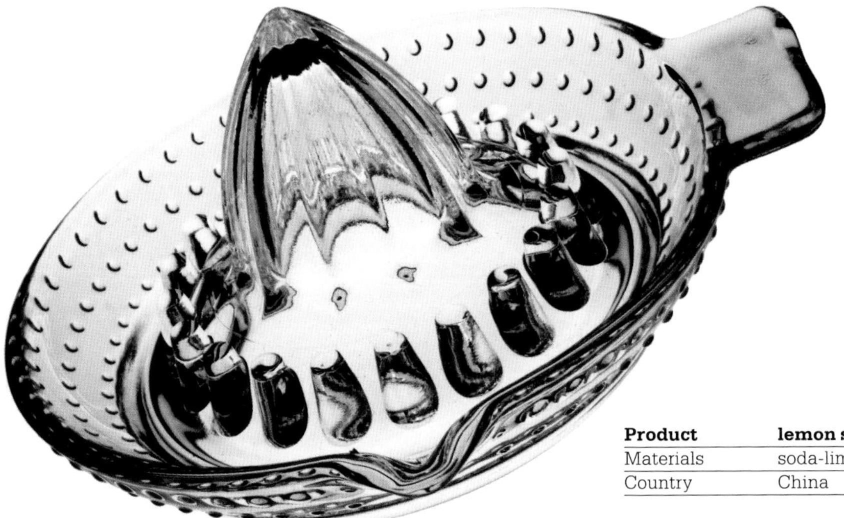

Product	**lemon squeezer**
Materials	soda-lime glass
Country	China

Along with pub ashtrays, this lemon squeezer – bought from my local supermarket – illustrates the complex, thick and chunky-walled forms that can be achieved by machine pressing glass in contrast to thinner, hollow, machine-blown glass products.

The thick-walled, chunky products that this process tends to produce are much more utilitarian than fine quality cut glass, which goes through a secondary process of grinding to achieve the crisp, sharp edges that are its hallmark. As in the case of other processes that generate products with a strong character, the particular 'look' and 'feel' of pressed glass has led to some pieces becoming collector's items.

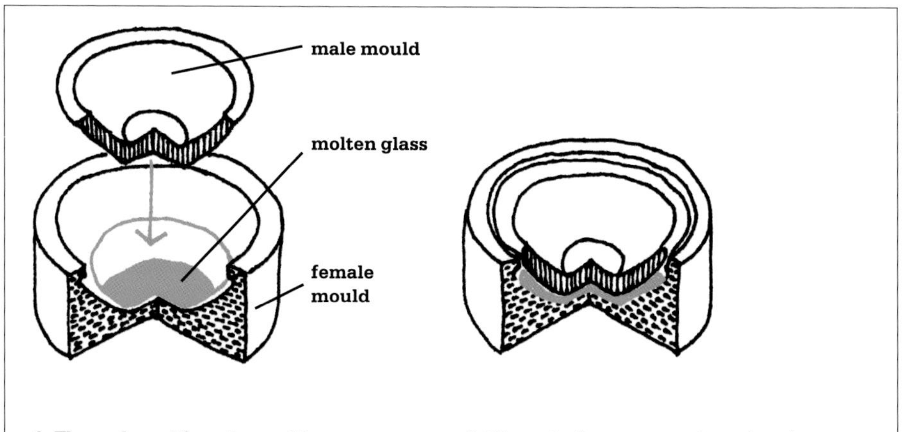

1 The male and female moulds are preheated and maintained at a steady temperature to ensure that the hot glass will not stick to the moulds.

2 The gob of gummy, molten glass is squashed between the two moulds. The thickness of the final component is determined by the amount of space left between the male and female moulds.

+

- Definition can be achieved on both the inner and outer surfaces.
- Allows for surface detailing that might not be possible with blowing.

–

- Its main disadvantage compared with blown-glass products (see p.98) is that it does not permit closed container shapes to be produced.
- Not suitable for making thin-walled sections.
- Generally involves more expensive tooling than that used in the mass-production of blown glass.

Volumes of production
Glass pressing is a term that can be applied to a hand process, a semi-automated or an automated machine process. Semi-automated production can be used for a minimum of 500 units, and is often employed for sampling large production runs for fully automated production.

Unit price vs capital investment
In fully automated production, the unit prices can be extremely low, but like most high-volume production, it requires expensive tooling.

Speed
In an automated set-up, and depending on the component size, a single machine can be set up to hold several moulds at the same time. This can result in huge production rates – some approaching 5,000 pieces per hour.

Surface
Dimples, serrations and diamond patterns are all achievable in pressed glass, although the definition is less pronounced than such patterns in cut glass.

Types/complexity of shape
While blown glass (see p.116) lends itself well to rounded shapes, pressed glass is a lot more versatile because it allows for more complex detailing and decoration. One of the key design features to bear in mind is that it is not possible to have a closed shape and, as in thermoforming (see p.64), the component must have a draft angle to allow for the mould to open at the end of production. Pressed glass is also more suited to thick-walled hollow ware.

Scale
Some semi-automated production allows for a maximum of about 600 millimetres in diameter. Larger pieces can be made, depending on the volume of production and the manufacturer.

Tolerances
Due to contraction and expansion of the material, glass pressing is able to match the high tolerance of engineered components. However, a typical tolerance is ± 1 millimetre.

Relevant materials
Almost any type of glass.

Typical products
Lemon squeezers, railway signal lamps, lenses, street and display lighting, laboratory glass, pub ashtrays, pavement lenses, wall blocks, marine and ship lighting, aircraft and airport runway lenses, and road and traffic signals.

Similar methods
Cut glass can be used for fine-detailed patterns, but this is really your best bet for producing open glass shapes with the potential for decoration on both sides. For plastic components, you might want to consider compression moulding (p.174).

Sustainability issues
Glass is a renewable material that is widely recyclable. Recycling it can reduce waste and virgin material consumption, and it retains its excellent clarity and appearance after reprocessing. However, its production and manufacture is less environmentally friendly as pressing glass requires various stages of intense heating which consumes large quantities of energy. Some harmful air-polluting compounds and particles are released during processing.

Further information
www.nazeing-glass.com
www.glasspac.com
www.britglass.org.uk

Pressure-Assisted Slip Casting

with pressure-assisted drain casting

Pressure-assisted slip casting is a development of conventional ceramic slip casting (see p.140). Compared with the traditional form, it offers several manufacturing advantages that affect the speed and complexity of the final component. Conventional slip casting involves the use of plaster moulds into which the ceramic 'slip' is poured. The 'de-watering' of this slip is based on a capillary action that draws water from the slip into the plaster, leaving the clay to form a dry layer against the internal wall of the mould. This can be quite slow and the plaster moulds have a limited life.

In pressure-assisted slip casting, a more resilient material, with larger holes, is used for the mould. The size of the holes means that the capillary action is reduced and replaced by the use of pressure (typically between 10 and 30 bar, depending on the size of the product). This involves pumping the slip into the porous plastic mould. Under this pressure, the water seeps out through naturally occurring capillary tubes in the mould. Once dried, the form is taken out of the mould and any imperfections are cleaned off. The product is then dried in fast dryers and sprayed with a glaze before firing.

A project called Flexiform, led by Ceram Research in the UK, has enhanced pressure-assisted slip casting, coming up with a process it calls 'pressure-assisted drain casting'. In this development, the conventional synthetic mould is replaced by a machinable plastic, which can be machined directly from the product designer's original CAD drawing. This offers a number of further advantages, including cheaper tooling and the possibility of the mould being re-cut, which is not possible with the moulds used for pressure-assisted slip casting.

Product	**bath from the Loo range**
Designer	Marc Newson
Materials	ceramic
Manufacturer	Ideal Standard
Country	UK
Date	2003

This bath is a typical example of the scale of casting that can be produced in ceramic.

Volumes of production
Pressure-assisted slip casting moulds typically require volumes of approximately 10,000 pieces to justify the use of the plastic tooling.

Unit price vs capital investment
Cost-effective unit parts, which are the result of several factors, outlined above. In the Flexiform pressure-assisted drain-casting project, mould costs are significantly reduced.

Speed
Conventional slip casting (p.140) can require anything up to an hour for casting, de-moulding and drying. Pressure-assisted slip castings can typically result in a reduction in time of 30 per cent.

Surface
Superior quality finish compared with conventional slip casting, with reduced casting seams resulting in less felting than with the conventional method.

Types/complexity of shape
From small and simple to large and complex parts with undercuts. Anything, from bathroom products to art objects and dinnerware, can be made using this process. Just think of the U-bends on the underside of a toilet to glean an understanding of the types of complexity possible.

Scale
From small teacups to toilets and baths.

Tolerances
As is the case with any fired piece, moulds need to be made that take account of a reduction in size once the product has been fired.

Relevant materials
Suited to most types of ceramic material.

Typical products
Complex tableware, which can require four-part moulds for teapots and coffee pots with integral handles. Apart from its large-scale use in sanitary ware, it is in the area of advanced ceramic technology that pressure-assisted slip casting is attracting the strongest interest.

Similar methods
Ceramic slip casting (p.140) and compression moulding (p.174).

Sustainability issues
Water is used to aid the ceramic flow instead of the organic solvents or binders that were previously used. This water can be cleaned and recycled back into the process to reduce waste and consumption of raw materials. Further material reductions are made through the increased durability of the plastic moulds, which can withstand extensive use.

Further information
www.ceramfed.co.uk
www.cerameunie.net
www.ceram.com
www.ideal-standard.co.uk

- **The plastic mould allows higher pressure to be used in the production of large pieces.**
- **Plastic moulds have a longer life (approximately 10,000 casts) before they are thrown away.**
- **Fewer moulds are needed and less storage is required.**

- **The moulds add to the set-up costs (however, for drain casting, Flexiform moulds greatly reduce the tooling costs).**

Viscous Plastic Processing (VPP)

As the technology behind materials and manufacturing techniques progresses, the previously barren spaces between different families of materials are bridged. Of all the material families, plastics make up the group that is the most versatile in terms of production techniques available. However, other materials, such as metals and ceramics, are all being explored to find new ways of mass-producing components using plastic-state forming techniques. This allows materials that have traditionally limited means of forming, including ceramics, to be formed using methods such as injection moulding (see p.196).

The material and the method of production go hand in hand in the sense that the properties of the materials dictate the complexity of production that is available. One of the problems in forming ceramics involves the need to eliminate the inherent microstructural defects in ceramic materials. These defects reduce the strength of the material, making it

Product	**teacup from the Old Roses range**
Designer	Harold Holdcroft
Materials	bone china ceramic
Manufacturer	Royal Doulton
Country	UK
Date	1962

VPP technology was used to enhance the properties of bone china, so that these uniquely British teacups could be injection moulded. A combination of design and the cost-effectiveness of this process means that 100,000,000 cups have been sold since 1962.

brittle. Viscous plastic processing, or VPP, is a method of enhancing the properties of ceramic materials that eliminates these flaws, resulting in a way of processing ceramics that is much more flexible and, to use the technical term, 'plastic' in its nature.

The process involves ceramic powders being mixed with a viscous polymer under high pressure. This mixture can then be used to form components through a range of fabrication techniques, including extrusion (see p.96) and injection moulding.

Volumes of production
Not applicable.

Unit price vs capital investment
Not applicable.

Speed
Not applicable.

Surface
An excellent surface can be achieved, depending on the grain size of the ceramic powder.

Types/complexity of shape
Because of the enhanced 'viscous-elastic behaviour' of ceramics produced in this manner, components have high strength in their 'green' state, which enables quite adventurous forms to be produced. The process also allows thinner wall sections to be produced than is the case with standard ceramic materials, which ultimately leads to higher strength parts with reduced weight.

Scale
It is possible to create large products, but not in all dimensions. VPP is, in other words, capable of producing long, extruded sections with wall thicknesses of up to 6 millimetres, or thin sheets.

Tolerances
Not applicable.

Relevant materials
Any ceramic material.

Typical products
Flat components, substrates for electrical components, kiln furniture, springs, rods and tubes, strength in green-state cups, body armour and biomedical applications.

Similar methods
Not applicable.

Sustainability issues
By enhancing the strength of the ceramic, thinner-walled parts can be produced to reduce material consumption and extend the lifespan of the product. This increased strength can also help to reduce possible defects during forming, which in turn helps to minimise waste of materials and additional processing. Any type of fabrication used to form the ceramic–polymer mixture requires extensive heat, which is energy intensive.

Further information
www.ceram.com

- **The process can be applied to a wide range of ceramic materials and offers materials with good 'green' strength.**

- **Limited number of manufacturers.**

7: Advan

Advanced and new technologies

The starting point for most of the processes featured in this section is that the information used to make the shape is supplied by a CAD file. This eliminates tooling costs, as do Smart Mandrels™, also featured in this section (though these are not driven by CAD), and together they all provide a complete mind shift from existing rules of production. On this basis, the methods in this section point the way to future industrial production and hint at the fact that these new technologies will provoke the biggest change in the nature of mass-produced objects since the Industrial Revolution. It is a group of processes that includes the relatively familiar process of stereolithography, but also has some new technologies that put manufacturing into the hands of the consumer.

Inkjet Printing

Desktop printers have allowed anyone with a computer to turn a desk into a place where all sorts of things can happen. The seemingly humble printer may well be the hub of a revolution that will change the way we make objects. The day will soon come when we will be able to download plans for a product (a door handle, for example) and make it from our own desktop three-dimensional printer, which has been loaded with the appropriate raw materials, in the same way that you load up your bread-maker last thing at night so that you can enjoy a fresh loaf in the morning. Before such three-dimensional technology becomes a reality at a domestic level, however, 'techies' are busy pushing the envelope to discover new applications for this familiar object, with its clanking robotics.

Already, Homaro Cantu, a chef based at Moto's restaurant in Chicago, has turned a Canon i560 inkjet printer into a machine for making food. Having replaced the ink cartridges, he prints edible liquids instead of CMYK inks onto an edible starch-based paper. In a move worthy of Willy Wonka (let's not forget the edible sugary grass and flowers in his chocolate factory), Cantu has abducted a printing process to create an entirely new concept in how you order – and what you can eat – in a restaurant.

Possibly one of the most unusual adaptations of this technology is one that has been developed by various teams of scientists across the world, who use 'modified' inkjet printers to build up living tissue. Based on the long-held knowledge that, when placed next to each other, cells will weld together, the process involves tissue being built up, using a thermo-reversible gel as a kind of scaffolding over each cell. The team

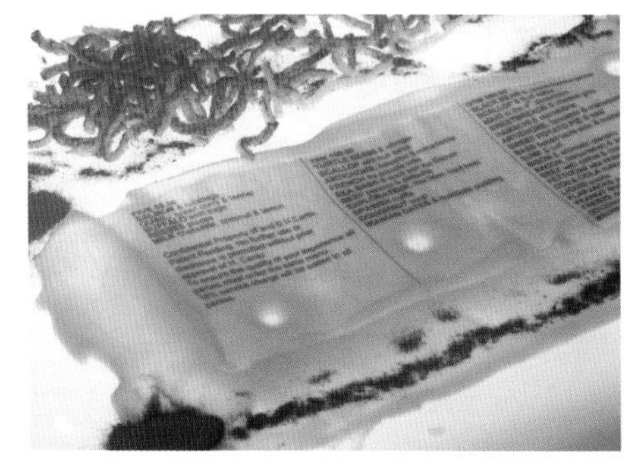

Product	**edible menu**
Designer	Homaro Cantu
Materials	vegetable-based dyes on edible paper
Manufacturer	Moto Restaurant, Chicago
Country	USA
Date	2003

This printed edible menu provides an example of an interesting crossover between the food and the production industry and shows that even on a 'techno' level food is providing a rich source of experiments.

that developed this, from the Medical University of South Carolina, uses the thermo-reversible gel as a way to support the cells as they are being distributed through the 'printing' action. This gel is interesting in itself, since it is designed to change instantly from liquid to gel (and back again) in response to a stimulus such as change in temperature.

Volumes of production
From one-offs to small batch production.

Unit price vs capital investment
Two-dimensional printers are within most people's budget, so you can take one apart and play with it at will, substituting the inks with anything you care to try.

Speed
Depends on what you want to do, but typically this is still a fairly slow process.

Surface
When making three-dimensional objects from standard production materials, the surface may have a ribbed texture as witness to the way the material has been laid down.

Types/complexity of shape
Highly complex shapes, restricted only by what you draw on your computer.

Scale
The team from the Medical University of South Carolina has demonstrated the highly controlled, cell-by-cell scale that is possible.

Tolerances
The production of three-dimensional living tissue demonstrates the fine tolerances that are achievable.

Relevant materials
Again, the machine is there to be explored, though you will need a basic combination of liquid and solid materials. The examples mentioned above give you some idea of the potential.

Typical products
The beauty with this process is that the examples mentioned above are currently a sort of DIY production, based on groups of people tinkering with technology and machines to give them new functions. The two contrasting examples illustrate that there is no such thing as a typical product for this hybrid technology.

Similar methods
Contour crafting (p.244), selective laser sintering (SLS) (p.252) and electroforming for micro-moulds (p.250).

Sustainability issues
The edible outcome of inkjet printing shown here is a particularly poetic example of how waste could be eliminated – the sheets can simply be eaten! Although experimentation is encouraged, care should be taken not to create too much waste or break too many printers through testing materials. In conventional ink-based printing the main issues are recycling and reuse of the cartridges.

Further information
www.motorestaurant.com

- **Allows for any shape generated on a computer to be turned into a three-dimensional object.**
- **Open to experimentation.**

–

- **Still in its infancy.**
- **Slow.**

Paper-Based Rapid Prototyping

Layered paper

The machine used for paper-based rapid prototyping makes the common inkjet printer look prehistoric and allows users to do extraordinary things. It is able to take humble, everyday A4 printer paper and create extremely detailed and intricate models of virtually any shape imaginable.

It allows you to take a drawing or scan from your computer and print it out into a 3-D physical object made solely out of sheets of paper. To do this the machine uses software that takes the drawing or scan and breaks it up into layers the same thickness as the paper. When the information is sent to the printer, it cuts each slice of paper to shape and layers them one on top of the other using a water-based adhesive. The layering takes several hours, but the end result is an incredibly precise 3-D shape made of hundreds of pieces of paper layered together.

Because of the flexibility of paper, a working, live hinge can be produced in one piece – something that cannot be achieved with many plastic-based alternatives – which means you get a more accurate prototype that is as close as possible to the real thing.

As only paper and a water-based adhesive are used, prototypes can be recycled, which makes this the most eco-friendly process on the market. What's more, recycled paper can be used in the first instance, with excellent results.

Product	**Mobile phone cover**
Materials	Photocopier paper
Manufacturer	Mcor Technologies Ltd
Country	UK
Date	Unknown

These mobile phone covers show the level of finish and detail that is achievable with this paper-based rapid prototyping process.

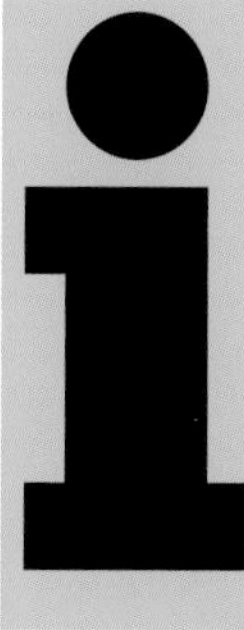

Volumes of production
As a form of rapid prototyping the process is ideal for low-volume production.

Unit price vs capital investment
The process is said to be up to 50 times cheaper than those used for plastics: paper costs very little and is available almost anywhere.

Speed
The mobile-phone covers shown here were built in 5–10 hours.

Surface
The surface has the consistency of wood and requires minimal finishing. The Z-axis resolution is 0.1 millimetres.

Types/complexity of shape
The process is suited to most complex shapes that can be produced by stereo-lithography. The exceptions are thin, spindly shapes.

Scale
The process is based on standard A4 paper and a maximum stack height of 150 millimetres.

Tolerances
Currently 0.1 millimetres on the XY-axis and 1 per cent on the Z-axis.

Relevant materials
Standard photocopier paper; Mcor Technologies recommend low-grade paper with a low fibre content for best results.

Typical products
The process is being developed for use in the medical industry, to reproduce X-rays in a physical form to help surgeons to plan ahead of operations, and in dentistry to enable orthodontists to create moulds of patients' teeth much more quickly than can be done using plaster. It is also expected that the process will be introduced for engineering, architectural and product design courses, for use by students as a cheaper alternative to traditional plastic rapid prototyping.
It is also a popular choice for architectural models, and as rapid tooling for moulds in processes such as vacuum forming, and also investment and sand casting.

Similar methods
Moulding paper pulp (p.149), pulp paper composite from Sodra Paper Labs (p.78).

Sustainability
Paper is a rapidly renewable resource, and is also widely recycled. No toxic fumes are released during processing, as they are with some plastic rapid prototyping, and components can be produced in a short timescale to make effective use of energy consumption.

Further information
www.mcortechnologies.com

+

- **Quick production time.**
- **Paper is cheap compared to alternative plastic resins and is readily available.**
- **More environmentally friendly than processes that use plastics.**

–

- **Currently limited to A4 size paper.**

Contour Crafting

This is a process that has the potential to revolutionise the construction industry. Dr Behrokh Khoshnevis, of the University of Southern California, has invented a machine that 'prints' houses. As he points out, the automation of the manufacturing industry has been advancing steadily ever since the Industrial Revolution. In comparison, developments in the construction industry have been meagre during the same period. However, this is something that Dr Khoshnevis plans to change with a process he calls 'contour crafting', an advanced form of spraying concrete.

Planned to be commercially available in 2008, the machines that are at the heart of this technology use a method of depositing concrete that is similar to that used in inkjet printers (see p.240) and extrusion (p.96). In this case, however, the technology is on a much larger scale, and it includes the ability of the 'printing' head to move in six axes and build up material in layers, based on CAD drawings rather than on two-dimensional graphics.

The 'printing' nozzles, which are suspended from an overhanging carriage, deposit quick-drying concrete, that is shaped by an integral trowel using a cylinder-and-piston system. A secondary feature of contour crafting is that the system allows utilities, such as conduits for electricity, plumbing and air-conditioning, to be embedded into the process.

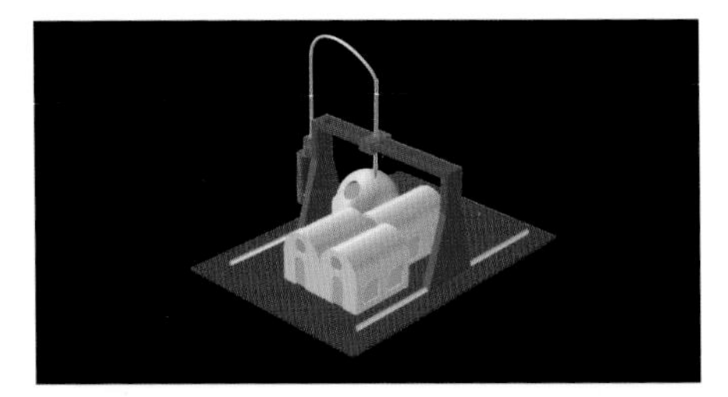

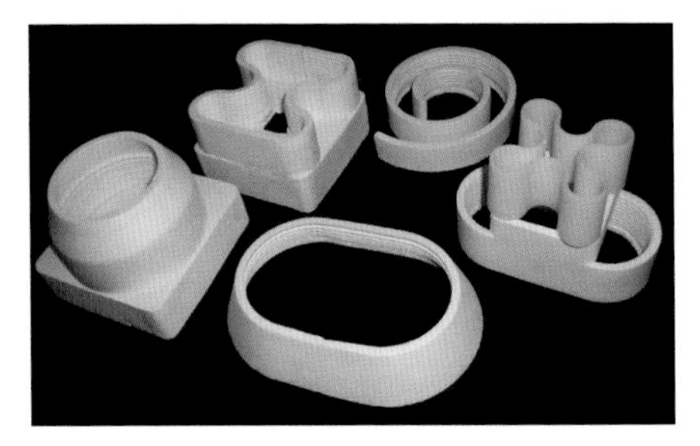

Product	**contour-crafted prototypes and CAD design drawing**
Designer	developed by Dr Behrokh Khoshnevis
Materials	concrete
Manufacturer	Dr Khoshnevis, under National Science Foundation and Office of Naval Research
Country	USA

These samples, though not on the scale of a building, illustrate the types of forms that can be created through contour crafting. Above is an example of the sort of CAD design that will drive the spraying process.

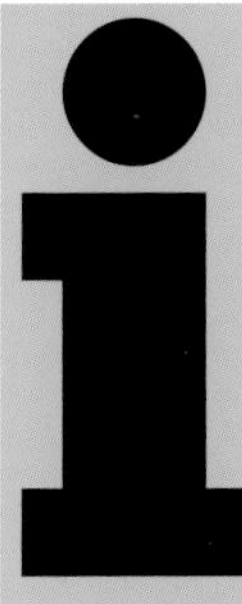

Volumes of production
The key feature of contour crafting is that it is an automated building method, however buildings can, of course, only be erected one at a time.

Unit price vs capital investment
Allowing for the fact that multiple houses can be built using a single machine, Dr Khoshnevis estimates the cost of building an average-sized American house at between a fifth and a quarter of the current cost of building a house by conventional means.

Speed
Construction using this process can build a 2,000 square-foot house, including electricity and plumbing, in less than 24 hours.

Surface
The use of the various types of trowel produces a good concrete surface, one that requires no preparation before painting. A painting system may even be incorporated within the contour crafting process itself.

Types/complexity of shape
The shape is limited only by the CAD drawing and the normal physical forces that apply to buildings, though even shapes such as arches can be extruded through the nozzle.

Scale
Dr Khoshnevis suggests that this method can be used for anything from a small house to a high-rise structure.

Tolerances
The nozzle assembly that can move in six axes allows for very high tolerances on a large scale.

Relevant materials
Cement, with additives such as fibre, sand and gravel.

Typical products
This is a process that is offering the building industry a new way to construct permanent houses, buildings and complexes, as well as temporary emergency shelters.

Similar methods
On this scale, the process is unique. The CAD-based system makes it similar to many smaller scale rapid-prototyping processes (see, for example, stereolithography (SLA) (p.246).

Sustainability issues
With its high-speed 'printing' system, contour crafting could dramatically reduce construction times and therefore energy consumption in comparison to traditional building techniques. There is no wasted or excess material as the concrete is built up accurately in layers to the exact contours of the structure.

Further information
www.contourcrafting.org
www.freeformconstruction.co.uk

- **Allows for rapid construction.**
- **Plans and designs can be easily altered because they are CAD-driven.**
- **It is possible to use local materials as reinforcement for the cement.**
- **Cost-effective.**
- **Automated process.**

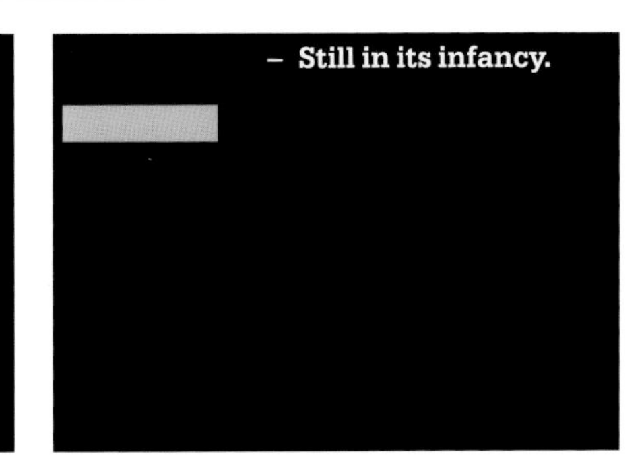

- **Still in its infancy.**

Stereolithography (SLA)

Stereolithography (SLA) is one of the best known methods of rapid prototyping. Driven by a CAD file, components are produced by a laser, which scans a bath of photosensitive resin, building the components layer by layer. The ultraviolet laser beam is focused onto the surface of the liquid, tracing the cross-section of the part and turning successive thin layers of the liquid into solid. The solid part remains below the surface of the resin throughout the process, because it is seated on a bed that is lowered gradually, allowing the component to be built up in layers.

All rapid prototyping technologies give a geometrical freedom that no other processes do. SLA is typical in that it allows for the

Product	**Black Honey bowl**
Designer	Arik Levy
Materials	epoxy
Manufacturer	Materialise
Country	The Netherlands
Date	2005

This beautiful, open-cell structure is an excellent example of the highly intricate and complex forms that can be built up using this process.

testing of components before entering into mass-production. Your choice of process is dependent on the geometry of the part, the surface quality required or the material that you want to use. Selective laser sintering (SLS) (see p.252), for example, cannot match SLA for quality.

SLA is an accurate process, although not the most accurate, and it can be applied to a range of materials, although not to as many as vacuum casting. (This is a method of producing small batches of identical components that are generally used for prototyping or modelmaking. It involves producing an original master that is cast into a silicone mould. The mould is subsequently filled with plastic resins. A vacuum is applied and the resulting parts are very accurate, with fine detail and thin wall sections.)

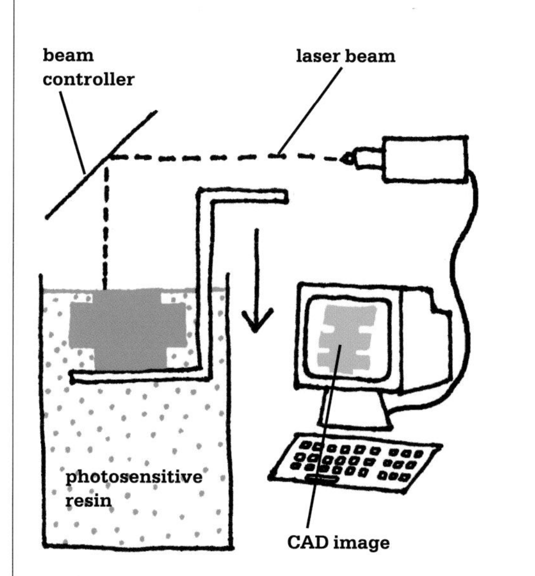

1 Driven by a CAD file, components are produced layer by layer by a laser scanning a bath of photosensitive resin.

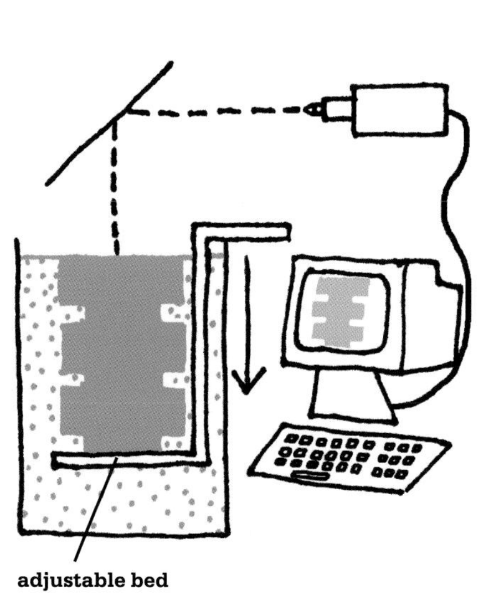

2 The ultraviolet laser beam is focused onto the surface of the liquid, tracing the cross-section of the part and turning successive thin layers of the liquid into solid. The part gradually sinks below the surface as it is lowered in a bed, allowing the whole structure to be built up.

1 This image, of designer Patrick Jouin's CI chair, shows the finished product being raised from the liquid polymer. During actual production the only part visible is the very top edge of the chair as it is formed by the laser.

2 The finished chair is seen this time with a white block that acts as an internal support for the seat during the forming process, without which the chair would collapse.

3 The completed chair before removal of the support block.

4 The finished chair in all its translucent, ghostlike glory.

+

- Unlimited geometric freedom.
- Good surface finish.
- No intermediate steps between the CAD model and finished object.

−

- High unit costs.
- Only photosensitive resins can be used.
- Inaccuracy in two directions.
- Often needs support structures.
- Not as rapid as many other prototyping processes.

Volumes of production
Due to the time it takes to build up a product, SLA is strictly limited to low-volume production.

Unit price vs capital investment
No tooling, and, even with a fairly high unit price, it is still the most cost-effective way of making prototypes.

Speed
Dependent on a number of factors, including the volume of the part, the material used, and the fineness of the step that is set by the operator. Another factor is the orientation of the component: if, for example, a drinks can is made lying down, the process is quicker, although less accurate, than when it is made standing up, which requires more passes with the laser.

Surface
The 'stepping effect' as a result of the layering can be controlled by the thickness of the step. Also, shallow gradients will produce lines similar to contour lines on maps. Steep gradients and vertical walls will have smoother surfaces, but in both cases the part may need sand blasting.

Types/complexity of shape
Anything that can be drawn on a computer.

Scale
Standard machines can allow for a 500 by 500 by 600-millimetre building area. For anything bigger than this the components must be made in several sections and joined together. However, some companies make their own machines, producing components several metres long.

Tolerances
Height is the least accurate dimension, because of the increased number of passes that the laser has to make, but tolerance is generally ±0.1 per cent plus 0.1 millimetre.

Relevant materials
Ceramic, plastic or rubber can be used. More commonly, engineering polymers such as acrylonitrile butadiene styrene (ABS), polypropylene and acrylic mimics are used.

Typical products
The word 'typical' has no application here, since you can make anything you want.

Similar methods
Vacuum casting (see above), selective laser sintering (SLS) (p.252) and inkjet technology (p.240).

Sustainability issues
Stereolithography requires a UV laser to cure the resin, and this is very energy intensive as cycle times can be quite slow depending on the complexity of the part. The additional support structures required for the majority of mouldings can increase material consumption and waste. However, the uncured liquid resin is washed off the finished part and can be recycled back into the process to help minimise material use. As with all rapid production methods, tooling is eliminated and in the future local production will eliminate transport costs.

Further information
www.crdm.co.uk
www.materialise.com
www.freedomofcreation.com

Electroforming for Micro-Moulds

Swiss company Mimotec has developed the process of electroforming (see p.164) to the extent that it can be used to make micro-moulds. Before describing the Mimotec process itself, however, I need to make it clear that micro-moulding is not the same as 'miniature' injection moulding. Micro-moulding is closer to the seriously minuscule nano-end of the scale, rather than just small-scale moulding, with parts being produced that can weigh as little as a few thousandths of a gram with details that measure only a few microns thick.

Although the principle behind micro-moulding is reasonably conventional, the methods used to produce the moulds are rather fascinating. Micro-moulds can be made by a number of different methods, including a micro-milling technique (where material is cut away). Mimotec, however, has harnessed the fine detailing achievable with electroforming to produce the most minute of moulds.

The Mimotec process starts with an unpolymerised layer of photo resist deposited on a glass plate. This is then exposed to ultraviolet light through a mask of the final shape, which causes the exposed resist to polymerise, leaving the non-exposed area to be washed away. The remaining part is coated with gold followed by a further layer of resist. The part is built up in this way to produce a more complex part, which acts as the moulding block and incorporates holes through which plastic for the component can be

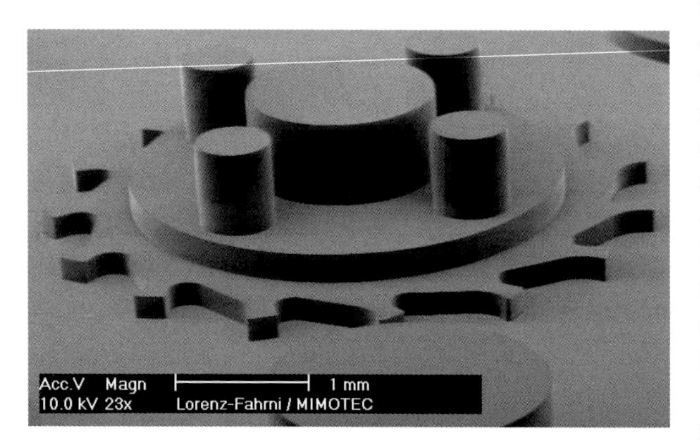

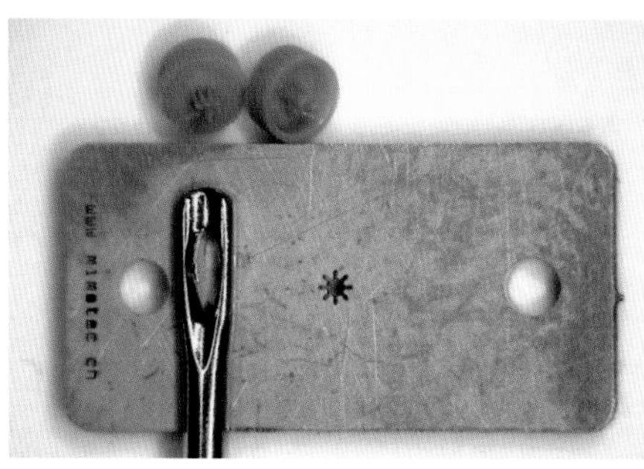

Product	micro-mould
Manufacturer	Mimotec
Country	Switzerland

A close-up image of the finished part (top) shows the scale achievable, as does the mould (beneath) that has a pinion cavity of only 0.6 millimetres and a micro-inscription on the side. The plate (as the presence of the needle demonstrates) is only 5 by 9.8 millimetres, and 1.2 millimetres thick.

injected. This process is just one of many new methods of forming nano-scale components, and it is an excellent demonstration of the ever-advancing research that is going on in this field of production engineering.

Volumes of production
Production runs of up to tens of thousands of components are possible using this type of micro-mould.

Unit price vs capital investment
The CAD-driven nature of this process means that the set-up costs are low.

Speed
It takes about seven hours to deposit a layer 100 microns thick, but several thousand micro-moulds can be made concurrently on a single glass plate.

Surface
It is possible to achieve high levels of detail and a fine finish on micromoulds made in this way.

Types/complexity of shape
It is not possible to make moulds that are capable of making shapes with tapered, or anything other than straight vertical, sides. Steps can be produced but require longer timings.

Scale
It is possible to create blocks of as little as 100 cubic microns, with embedded channels 30 microns wide. The largest parts are 100 by 50 millimetres.

Tolerances
± 2 microns.

Relevant materials
The micro-mould itself is made from gold with a nickel alloy coating. The moulded parts are generally made from polyacetals (POM) and acetal resins.

Typical products
As you might expect, the micro-moulds are used to produce very small parts for biomedical devices and electronics, watch making and telecommunications components.

Similar methods
Wire EDM (p.44) and micro-milling techniques.

Sustainability issues
Although the process can be very slow, the moulds produced using this method require no heat treatment or further processing such as polishing, which can significantly reduce energy consumption. As the component is built up in layers to the exact contours of the design, no cutting or machining away of materials is necessary, so the process makes very efficient use of resources and eliminates waste. The moulds have an above-average life expectancy to ensure continuous use.

Further information
www.mimotec.ch

- **Capable of extreme precision.**
- **Low set-up costs for electroforming make it good for prototyping.**

- **Making micro-moulds in this way is a fairly slow process.**
- **Restrictions in current technology mean that only nickel and phospho-nickel alloys can be used for the micro-moulds.**

Selective Laser Sintering (SLS)

with selective laser melting (SLM)

Innovation in production techniques has recently been dominated by advances in rapid prototyping. Designers are increasingly able to exploit the potential to make unique objects directly from a CAD file on a computer and selective laser sintering (SLS) is just one of the significant developments, opening up a world of rapid prototyping.

Sintering (see p.168) is a significant part of the field of powder metallurgy and it can be used in a number of different production methods. Selective laser sintering is an adapted (and refined) form of sintering in which a laser is used to solidify precise areas in a powder block in order to produce lightweight components. As in any sintering process, a powdered material (in the case of the implants illustrated here, a metal) acts as the starting point. A laser, driven by a CAD file, is fired repeatedly into the powder, fusing the particles together layer by layer until the specific component is built up. The process is also known as selective laser melting (SLM) for obvious reasons.

This, however, is only the beginning for the team at Renishaw PLC in the UK who use the technology to produce a type of microscopic scaffolding. They are able to exploit the design potential of a CAD file to produce components with a tiny, but complex, lattice-like structure. This results in forms that are made up mainly of air, like a sponge. The advantage of this type of micro-scaffolding is that it enables components to be produced in metals with a very high strength-to-weight ratio – the density of stainless steel parts, for instance, can be reduced by as much as 90 per cent compared with conventional processes.

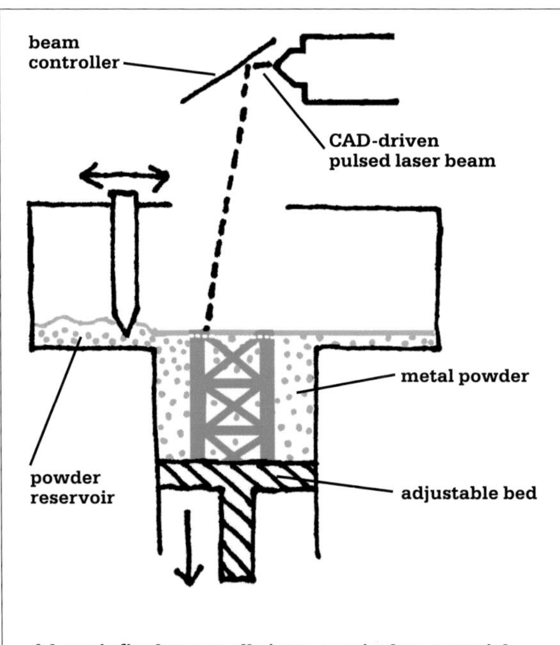

A laser is fired repeatedly into a particulate material, fusing the particles together at the point of impact until the specific component is built up, layer by layer.

Product	**sample of a hierarchical structure produced using SLM technology**
Designer	not applicable
Materials	stainless steel
Manufacturer	Renishaw PLC
Country	UK
Date	2005

This structure, only 3 centimetres high, was produced to demonstrate the small scale of work that is achievable using selective laser melting (SLM) technology.

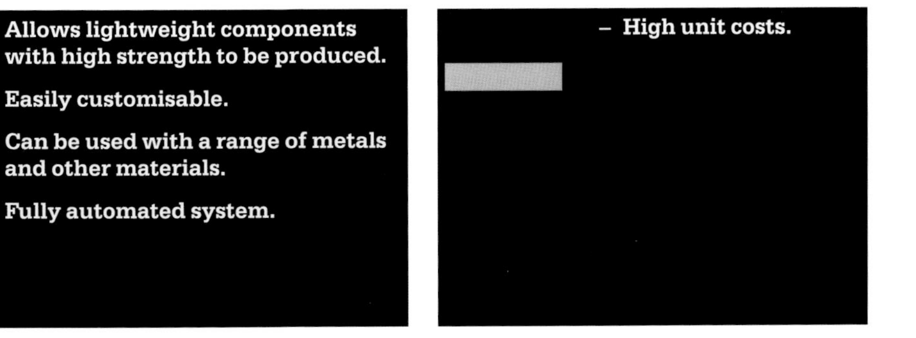

- **Allows lightweight components with high strength to be produced.**
- **Easily customisable.**
- **Can be used with a range of metals and other materials.**
- **Fully automated system.**

- **High unit costs.**

Volumes of production
Each component is made individually.

Unit price vs capital investment
No tooling, but the unit price is high because parts are individually made.

Speed
Although it is anticipated that selective laser sintering will eventually be taken up more widely for end production, it is still a fairly slow and low-volume process best suited to prototyping.

Surface
Components currently show a surface roughness of 20–30 microns – very little, in other words.

Types/complexity of shape
Limited only by the CAD technology that drives it, the microstructure of the Mining and Chemical Products components demonstrates that this is about as good as it gets when it comes to creating complex shapes.

Scale
It is possible to achieve very fine details, such as thin vertical walls with a thickness of as little as a tenth of a millimetre, while the overall size of parts is limited by the size of the powder block reservoir that the machine can hold.

Tolerances
Extremely high.

Relevant materials
Any particulate material used in powder metallurgy: metals, including steel and titanium, and plastics.

Typical products
SLS, principally a form of rapid prototyping, was initially used to test models before production. However, designers are pushing the technology towards production of finished products. This particular technology can be used to make anything from jewellery and heat sinks for computers to medical and dentistry implants.

Similar methods
Other CAD-driven technologies including deposition prototyping (contour crafting, p.244, for example), stereolithography (SLA) (p.246) and three-dimensional printing (such as adapted inkjet printing, p.240).

Sustainability issues
To make efficient use of materials and eliminate waste, components are built up in layers to the exact contours of the design thus eliminating secondary cutting or manufacture. The levels of complexity that can be achieved in these micro-structures also allow for reductions in material use and therefore weight. Because of the minute size of the components being made, several parts can be produced within the powder bed to increase productivity and energy efficiency. Unlike stereolithography, SLS does not require a support structure so waste and material consumption is lower. As with all rapid production methods, tooling is eliminated and in the future local production will eliminate transport costs.

Further information
www.renishaw.com

Smart Mandrels™ for Filament Winding

Shape-memory alloys and polymers are big news in the world of materials. Characterised by their ability to be 'programmed' to a particular shape, once heated and softened they can be bent out of this shape and re-formed into a new shape, which is retained once the material has cooled. The clever part is that when reheated, the part will return to its original 'programmed' shape.

The US-based company, Cornerstone Research Group, one of the major world players in shape-memory technology, has exploited such materials to develop a patented tooling system, Smart Mandrels™, for producing mandrels for the process of filament winding (see p.158). This system can be used in two ways.

In the first case, a single-shape memory mandrel can be formed into a specific shape, used to produce the relevant components, and then reheated, re-formed and reused to form a new shaped mandrel for an entirely different component. The second application is in the forming of a complex mandrel, one that might otherwise have been impossible to remove from inside the final component because of undercuts, and so on.

Filament winding using Smart Mandrels™ means that the filament can be wound around the mandrel, which is subsequently heated, softened and returned to its 'programmed' straight tube shape. This allows the completed filament winding to be easily removed.

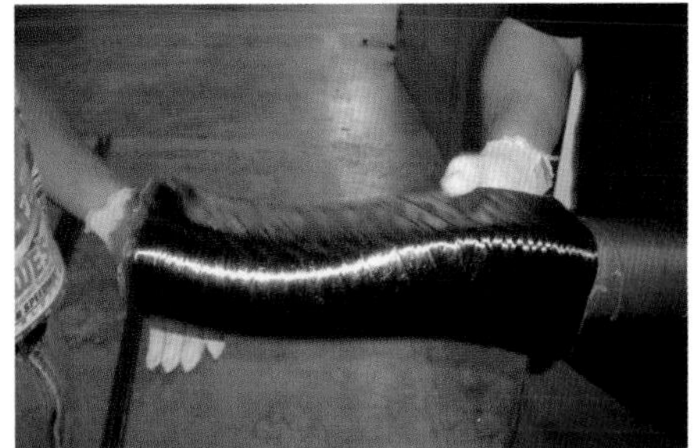

1 Winding onto the purple Smart Mandrel™ begins.

2 The Smart Mandrel™ is heated and softened for easy removal from the completed winding.

Volumes of production
For now, small runs and prototyping only, but this recently developed process will be equally suitable for large-scale production, since the mandrels are durable and can be used to make many parts.

Unit price vs capital investment
Smart Mandrels™ offer big savings for low production runs. This is because there is no need for expensive, multi-piece tooling, with the price staying at the same level for large production runs.

Speed
Cycle times are several minutes for each part, but it is significantly quicker than conventional filament winding with rigid mandrels (see p.158) because there is no need to assemble and disassemble the mandrel for each part.

Surface
No post finishing necessary, but the parts do have the distinctive 'look' of filament-wound products.

Types/complexity of shape
The main advantage with Smart Mandrels™ is that they allow for more complex forms to be produced using the filament-winding process. These can incorporate undercuts and returns that would normally be impossible to produce, because the mandrel could not be removed from the component.

Scale
Machines can be built to produce filament windings to a massive scale. The only limitations on scale will be the size at which the shape-memory alloys and polymers can be made and remain effective.

Tolerances
Not the kind of process that is suitable when high tolerances are required.

Relevant materials
Any thermoset plastic material, and glass or carbon fibre.

Typical products
Aeronautical components, tanks, rockets and housings.

Similar methods
Pultrusion (p.99), and contact moulding (hand or spray lay-up) (p.152).

Sustainability issues
Filament winding is largely automated so electrical energy is required to power the motors. The high speeds at which the machines can operate help to make efficient use of this energy through high-volume production. The high strength-to-weight ratio also offers significant weight savings.

Further
www.crgrp.net

- **Capable of producing highly versatile shapes.**
- **Reduced labour costs due to the ease with which the mandrel can be removed.**
- **Reusable and adaptable tooling.**
- **Simple to remove mandrel from component.**

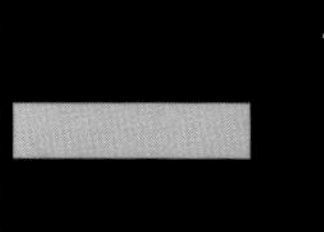

- **All parts have the distinctive 'look' of filament-wound products.**
- **Limited availability because it is a patented process.**

Incremental Sheet-Metal Forming

One of the major research areas in manufacturing at the moment is in the arena of 'industrial craft', a term that embraces a range of technologies that allow for a very flexible approach to mass-production by eliminating the need for specialised tooling. Incremental sheet-metal forming has the potential to revolutionise sheet-metal forming, making it available for low volumes of production for customised parts.

In essence, incremental sheet-metal forming is a type of rapid prototyping for sheet metal using a mobile indentor, so that almost any three-dimensional shell-shape can be made, without the need for specialised tooling. It is a term used to describe a number of methods of sheet forming that employ a generic, single-point tool that presses against a metal sheet in three axes (the work piece is held in a clamp), depressing it into a shape based on a path that is supplied by a CAD file.

The process has been in use for 15 years, but its potential is still not widely adopted in industry, chiefly as a result of the difficulty in assuring

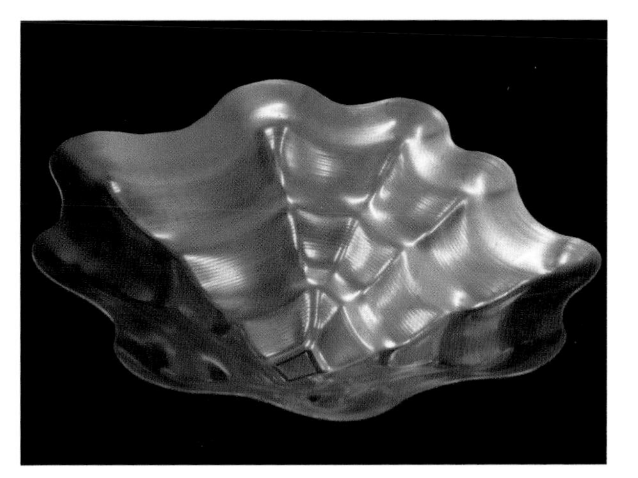

Product	**sample of incrementally formed sheet**
Materials	stainless steel
Manufacturer	sample produced by Institute for Manufacturing, Department of Engineering, University of Cambridge
Country	UK
Date	2006

Researchers Julian Allwood and Kathryn Jackson of Cambridge University are two of the many researchers internationally who are looking at ways of developing the process for wider industrial use. The stepping seen in this sample illustrates the path of the tool as it traces across the metal sheet slowly pushing it into shape.

geometrical precision in the formed part. However, Toyota has explored the process for forming parts for prototyping cars, using a one-sided die in order to gain more control.

There are a number of researchers exploring different variations of the process, some of whom are using two indentor tools at the same time, on either side of the work piece. Negative and positive dies can also be employed to give greater control of geometrical accuracy and surface finish.

This close-up image shows the single-point tool poised over the clamped sheet of metal, which is about to be formed into shape.

1 The shape of the component is drawn as a CAD file.

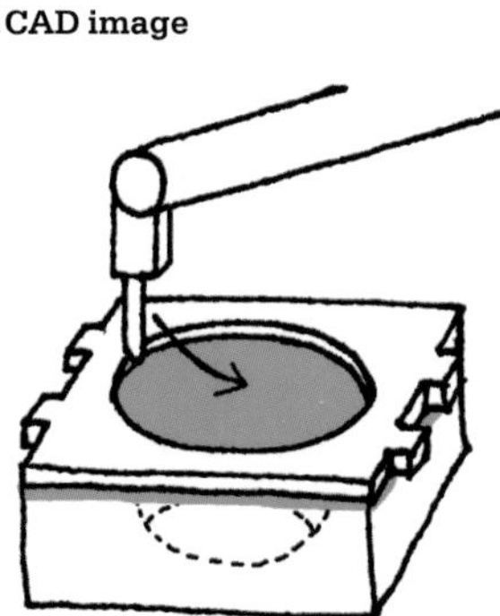

2 The metal sheet is fixed into a clamp and a single-point tool presses the sheet into shape.

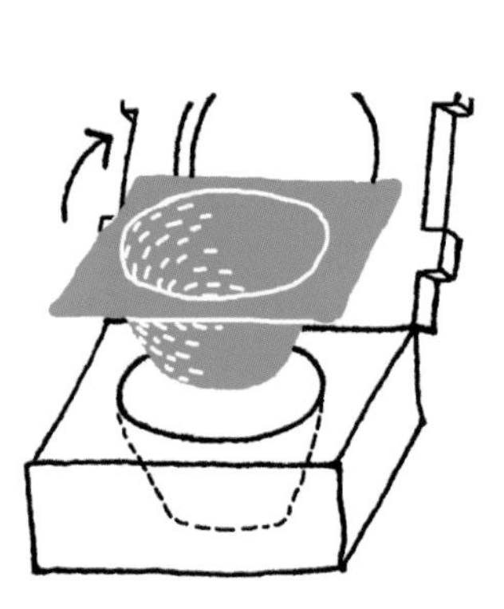

3 The final component is removed.

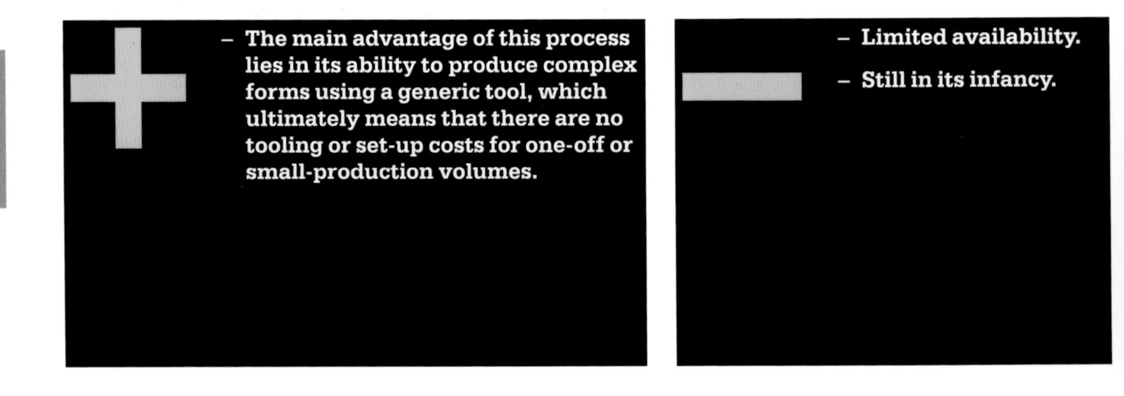

– The main advantage of this process lies in its ability to produce complex forms using a generic tool, which ultimately means that there are no tooling or set-up costs for one-off or small-production volumes.

– Limited availability.

– Still in its infancy.

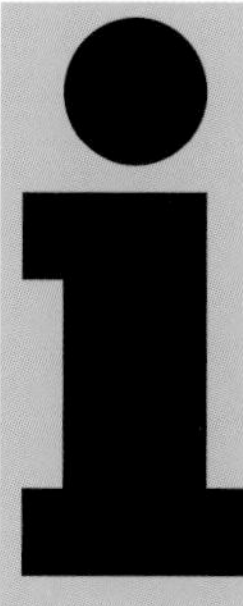

Volumes of production
The process of incremental sheet-metal forming is increasingly well known and is attractive because it offers the possibility of economic production of small batches. It has been used for manufacturing prototype products, including a prototype car made by Toyota. Other applications such as dental prosthetics, where each product must be unique, are also emerging.

Unit price vs capital investment
The obvious advantage of this process lies in the fact that it allows low-volume production with extremely low tooling and set-up costs.

Speed
Typical feed-rates can be up to 50 millimetres per second, and a typical part will take between 20 minutes and an hour, depending on the surface quality required.

Surface
Depends on the step size between successive passes of the tool. A step size of around 0.1 millimetre per pass gives an A-class surface, as rated by a car body maker. The surface can also be enhanced by the use of moulds.

Types/complexity of shape
Depends on whether or not a die is used, but the parts will always be shell-shaped – although, in the near future, machines will be built with upper and lower indentor tools to get around this.

Scale
While typical components are approximately 150–300 millimetres square with an average thickness of 1 millimetre, researchers in Japan are capable of forming parts that range from a few millimetres long up to sheets that are 2 metres long.

Tolerances
Depend on whether or not a die is used. First-time geometric accuracy can be poor (out by 2–5 millimetres), even if the tool path is only creating simple contours from a CAD model. This can be improved, but it involves trial and error. Accuracy is obviously much greater if a die is used.

Relevant materials
A wide range of materials, including a selection of aluminium and steel alloys, stainless steel, pure titanium, brass and copper.

Typical products
Several applications exploit the potential of this process for one-off production, including the manufacture and repair of car body panels, tailored medical devices and prosthetics, and architectural panels.

Similar methods
Incremental sheet-metal forming has its roots in metal spinning (p.56), but obviously has far greater advantages in terms of rapid prototyping and flexible manufacturing. Another closely linked process is press forming (see metal cutting, p.59).

Sustainability issues
The process eliminates the need for specialised tooling for each component produced, and dramatically reduces material use in mould manufacture and subsequent energy use. Faulty sheets can simply be reworked rather than reprocessed or scrapped, which further minimises energy use and also allows for easy modification of pre-formed parts. In terms of batch production and prototyping, processing speeds and energy use can be more efficient than in competitive techniques.

Further information
www.ifm.eng.cam.ac.uk/sustainability/projects

Finish Techni

In his visionary book *The Materials of Invention*, Ezio Manzini defines the surface of objects as 'the location of the points where an object's material ends and the surrounding ambient begins'.

The surface of a product can often be the simplest and most viable location for invention. In 2010 the appliance manufacturer Miele launched a special-edition vacuum cleaner that was covered in a peachy, flocked surface. It transformed a product that would have been made with a very ordinary high-gloss plastic into a completely different visual and tactile surface.

Including techniques such as painting, plating, and covering this section presents many of the standard and often widely available processing methods that fit into the areas of practical and decorative coatings. It also includes a snapshot of some of the increasing number of high-tech and smart coatings that are adding a new kind of functionality to products.

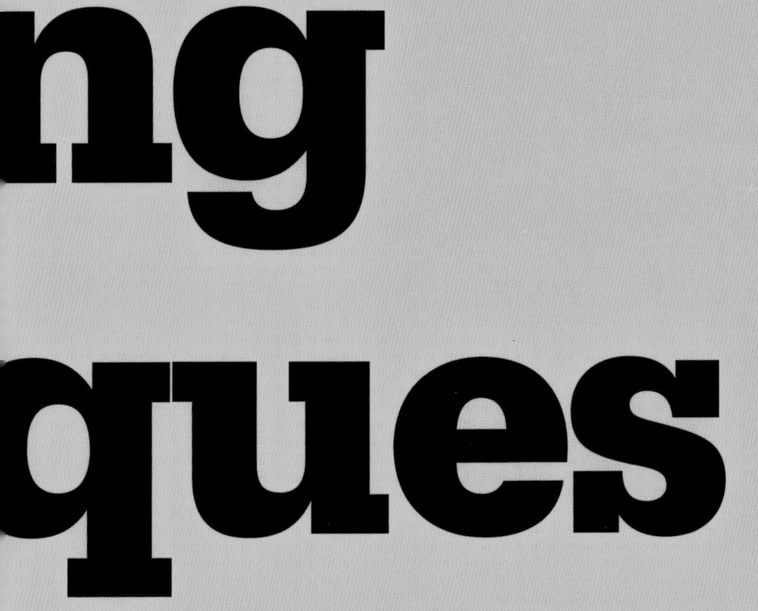
ng
ques

Decorative

Sublimation Dye Printing

Sublimation dye printing is used exclusively to decorate pre-moulded, three-dimensional plastic products. Colours, patterns and graphics, like those seen on Massimo Gardone and Luca Nichetto's Around the Roses tables (above), can be applied but the process offers no protection; instead it provides decoration that does not diminish even with scratches. Unlike silk-screening or painting, it is able to produce a full spectrum of colours, images and designs.

The particular type of dye that is used vaporises when heated, bonding to the molecules of the plastic substrate and absorbing the colour into the component down to a depth of 20–30 microns. As a result, the surface cannot wash or rub off, making a fully decorated product that is highly durable and scratch resistant. The technique is also used to create decorative surfaces for a variety of applications, including the lid of Sony's VAIO laptop computers. The computers are decorated with a range of different colours and graphics to give them personality and individuality.

Typical application

Sublimation dye printing is also used in photo-quality printers, where colours are printed as solid dyes which are heated up and permeate the paper before returning to solid form. This creates a much higher-quality image than dot-matrix printers such as inkjets and laser printers, and the prints are less vulnerable to fading and distortion.

Sustainability

The process is efficient and safe for the environment. However, a great deal of heat is produced and energy usage can therefore be an issue.

Further information

www.kolorfusion.com
www2.dupont.com

Vacuum Metallising

It's not exactly the real thing but if you need to give a metal effect on plastic components this is the process for you. It is difficult to chrome-plate plastics using conventional electroplating, but vacuum metallising is a cost-effective and widely used method of achieving similar results.

During the vacuum metallising process, aluminium is evaporated in a vacuum chamber, and then condenses and bonds to the substrate to form a chrome-like layer. A protective topcoat is then applied to the surface. The coating is much cheaper and more environmentally friendly than chrome plating, although it does not reach quite the same high level of durability and corrosion-resistance. Although the plastic component will not have the weight or coldness of real metal, it is worth considering for parts that will not be touched by the user's hands; this might help to carry the illusion off more easily. An application that uses the process for enhanced illumination is in the cones for torches. Tom Dixon applied it to a different form of illumination in his copper shade made from polycarbonate plastic with a metallised copper finish (above, www.tomdixon.net).

Typical application

The conical flashlight reflectors at the end of every torch and car headlamp, and automotive trim with a chrome-like shine, are both products that use vacuum metallising.

Sustainability

Vacuum metallising uses aluminium to create a chrome-like effect, which is much more environmentally friendly than chrome plating.

Further information

www.muellercorp.com/chromeplatingplastic.htm

Flocking

Flocking is a surface that has very strong assocations: fuzzy-felt and my mother's wallpaper in the 1970s, and the repellent feel of scraping fingernails over that tight, furry surface. It's a technique that was traditionally used for decorative purposes, but also has many other practical advantages, such as sound and heat insulation, which make it ideal for a wide variety of applications. An unusual example is the special edition Miele 56 vacuum cleaner (above).

Flocking involves applying precision-cut lengths of fibres to an adhesive-coated surface, using an electrostatic charge. This creates a seamless fabric-like coating, with up to 23,250 fibres per square centimetre. The length and type of fibre used determines the type of finish produced.

Typical application
Flocking is widely considered to be simply decorative, but it has a number of advantages that make it suitable for a wide variety of purposes. For example, it is commonly found in the interiors of cases for spectacles, jewellery, cosmetics and so on where protection is required. Flocking can also reduce condensation, so is often used for caravans, boats and air-conditioning systems. Two of the most innovative design applications are coating ceramic tableware with a flock surface and the special-edition vacuum cleaner featured here.

Sustainability
Any excess flocking that does not attach to the surface can be collected and reused. Products which have been flocked can be recycled, depending on the type of fibre and base material used.

Further information
www.krekelbergflockproducts.nl

Acid Etching

Also known as chemical milling or wet etching, acid etching is great for producing intricate patterns on thin, flat metal sheets. It involves a resist being printed onto the surface of the material to be treated. The resist consists of a protective layer that is able to withstand the corrosive action of the acid, which therefore eats away only the exposed metal. The resist can be applied in the form of a linear pattern, a photographic image or any combination of the two.

Typical application
Acid etching is often used in precision electronic components such as switch contacts, actuators and microscreens, and can also be used for labels and signs. Designer Tord Boontje used the process to produce his Wednesday Light (above), a lampshade that folds out from incredibly detailed sheets of etched stainless steel. Acid etching is used in the military to make a flexible trigger device on missiles, which is so sensitive it bends according to air pressure the closer the missile gets to its target.

Sustainability
Many metals are recyclable and the use of toxic chemicals is minimised in modern forms of acid etching, which gives the process a good environmental rating.

Further information
www.precisionmicro.com

Laser Engraving

You probably have some idea of how engraving works, and are familiar with its use for lettering on objects like trophies and plaques. However, laser engraving can produce microscopic details that are so accurate and precise that it can be used to produce printing plates for banknotes that are difficult to counterfeit. Although there are various types of engraving, it is now usually done with a laser, as this is particularly effective for mass-production and intricate detailing.

The engraving machine uses a laser that acts like a pencil – the beam is computer-controlled to trace patterns on the surface of the material. Direction, speed, depth and size can all be fine-tuned to suit the application. Laser engraving does not use tool bits or contact the surface in any way, so parts do not need to be regularly replaced as they do with hand engraving.

Typical application

Some of the highest quality hand engraving can be found on jewellery, but jewellers have realised that by using a laser they are able to engrave with even greater precision and at a much greater speed. Because the laser can cut into both flat and curved surfaces, the process has become particularly effective in this area.

Laser engraving has another, slightly more unusual application – in architectural models, where it can be used to create very fine details and patterns.

Sustainability

Unlike many other surface decoration processes, laser engraving does not involve consumables or problems with toxic by-products. However, some materials do emit hazardous gases when they are laser-cut.

Further information

www.norcorp.com
www.csprocessing.co.uk

Screen-Printing

Screen-printing is arguably the most versatile of all printing techniques. It can be used with many different types of material, including textiles, ceramics, glass, plastic, paper and metal, and can also be used to print onto objects of any shape, thickness and size – making it ideal for a broad range of applications.

The process relies on a woven mesh screen which is stretched tightly over a frame. The graphic pattern is produced on the screen by masking off the negative spaces, either manually or using a photochemical process. Ink is forced through the threads of woven mesh and onto the open areas of a stencil using a roller or squeegee. This produces a sharp-edged shape on the surface. The type of ink used, diameter of the threads and thread-count of the mesh all affect the final image.

Typical application

Screen-printing is most commonly associated with clothing, but the technique is used for many other applications, including clock and watch faces. More excitingly, it is now being utilised for more advanced uses like laying down conductors and resistors in electrical circuits that are on top of ceramic materials.

A rotary screen-printer is used to speed up the process of printing onto T-shirts and other garments. The clothing industry accounts for over half the screen-printing done in the US.

Sustainability

The screen can be reused once it has been cleaned. Screen-printing can produce prints at a much quicker rate than comparable methods, making it more efficient in terms of energy consumption.

Further information

www.fingerprint-comms.co.uk

Electropolishing

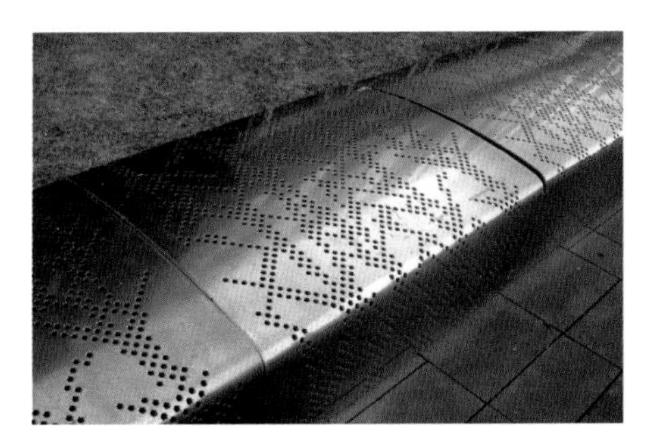

Many metals that look and feel smooth to the touch are, in fact,the opposite. Examining a surface under a microscope can reveal that it is full of tiny imperfections, which may affect how well the metal performs in use. This is where electropolishing comes in. It uses an electrochemical process to remove a thin layer of the metal, to expose a cleaner, brighter and smoother surface.

Electropolishing is in effect the opposite of electroplating, where material is added to a surface. The process begins with immersing the work piece in an electrolyte bath through which an electric current is passed. This starts an oxidation process, that causes the surface of the metal to dissolve. Increasing the processing time removes more metal. The image above shows the process applied to the Berta Vilagrassa bench by KX Designers.

Typical application

Electropolishing removes all traces of hydrogen from the surface, which greatly restricts the growth of bacteria. For this reason, it is commonly used in the food industry for food-processing and food-handling equipment.

The process is ideal for small and complex products made from alloys such as copper; almost any other finishing method will damage such soft metals.

Sustainability

Although electropolishing involves the use of potentially harmful chemicals, the electrolyte bath can be used over and over again for a number of components, with a minimal amount of waste.

Further information

www.willowchem.co.uk

Tampo Printing

Tampo printing is a versatile technology that can be used on virtually any material or surface. It is a very effective way of creating high-standard graphics or other decoration on complex parts, as it allows the graphic to be printed onto small, confined surfaces and curves. However, only solid colours with no gradients or tints can be produced.

The process begins with the graphic being produced on a film, to exact size. It is then chemically etched onto the surface of an anodised plate. The plate is positioned in the tampo printing machine where ink is distributed across the surface. The plate is then scraped clean so that the only ink is within the etched graphic. A silicone pad is lowered onto the metal plate and the ink of the graphic is picked up. The silicone pad then moves and presses the graphic onto the surface of the item being printed.

Typical application

Tampo printing is the main method used to print logos on general promotional merchandise such as pens and key rings. It is also useful for such products as calculators, radios, clocks and torches.

Sustainability

The tampo printing machine is CNC-controlled, using lasers to set up parts accurately and quickly, which makes the process energy efficient.

Further information

http://www.aki.co.uk/page/tampo_pad_printing

Suede Coating

Think of a peach skin or a fine flocked surface, then add a subtle rubbery feel and you have what was the material of the moment in the 1990s. Nextel®, a surface coating with the velvety feel of the skin of a peach, was developed at the request of NASA, who needed a coating with some specific and demanding properties. The initial use for Nextel® was interior surfaces on space shuttles, which need to be anti-static, chemically inert and non-reflective, as well as scratch resistant. However, the coating is now used in all kinds of industries for both decoration and function.

Nextel® is very easy to apply and consists of neoprene granules in a special carrier medium. The coating consists of three layers. The first is the substrate, which is coated with the second layer – a suitable primer. Once this has dried, colour is applied and tiny neoprene granules, which make up the third layer of the coating, are then added. Standard industrial spraying equipment is used and the coating is either air-dried, or alternatively, dried in a low-temperature oven.

Typical application

Applications for Nextel® are almost unlimited, but the range of materials that can be coated makes it particularly suitable for interior design. Its toughness and aesthetic properties mean it is ideal for furniture for both office and domestic furniture.

The coating has also been widely used in the field of transportation. Car dashboards, and seating for trains and aeroplanes, are just a few of the applications that make use of this hard-wearing and soft-surface finish.

Sustainability

Nextel® conceals imperfections such as dimples on the surface of the substrate. This reduces the number of processes that are required and therefore the amount of energy used. Little or no waste is produced during the application of the coating.

Further information

www.nextel-coating.com

Hot Foil Blocking

Hot foil blocking is a 'dry' process, which means no inks or solvents are involved and the work can be handled straight after it has been blocked. It is versatile and can be used on a whole range of materials. The result is a decorative effect that has a brilliance that cannot be produced using inks.

To begin the process, a metal blocking die or plate is produced by etching the graphic into the metal so that it is raised. The die is placed in the blocking machine, which consists of a roll of foil and a shelf beneath the foil where the material to be printed is placed. When the die is pressed down onto the foil, the heat and pressure from the die cause the pigment in the foil to be released and this is transferred onto the material in the shape of the raised image on the die. Hot foil blocking is often combined with embossing to make graphics stand out even more.

Typical application

Typical applications for hot foil blocking include book covers, business cards, toys and premium packaging. The process can also be used for some more unusual applications, such as labelling shoes and the holographic detail on credit cards.

Sustainability

Hot foil blocking does not use a lot of energy, although there is inevitably some waste material.

Further information

www.glossbrook.com

Over-Moulding

Over-mould decoration is often seen as an extension of injection moulding, rather than a manufacturing process in its own right. It is, however, an incredibly useful tool that enables designers to add almost craft-like qualities to products because of the way it allows for different materials to be combined.

Although it is capable of producing very complex products, the process is in itself fairly straightforward. The main part is moulded before being transferred to the next mould, where a second material is moulded around, over, under or through the base moulding.

Typical application
Over-moulding is often used for products that fall into the category of personal mobile technology, including mobile phones, PDAs and laptops.

A number of mobile phone cases have a small patch of fabric on their surface. You would imagine that adding the fabric would require a whole new process, but over-moulding allows for plastics to be combined with other materials during the actual moulding process, which eliminates the need for any secondary finishing.

Sustainability
It can be difficult to recycle over-moulded products, as they consist of a combination of materials which, in many cases, aren't easily separated. This is largely a problem for designers, who need to bear this in mind and make sure the product can be dismantled effectively.

Further information
www.ecelectronics.co.uk

Sandblasting

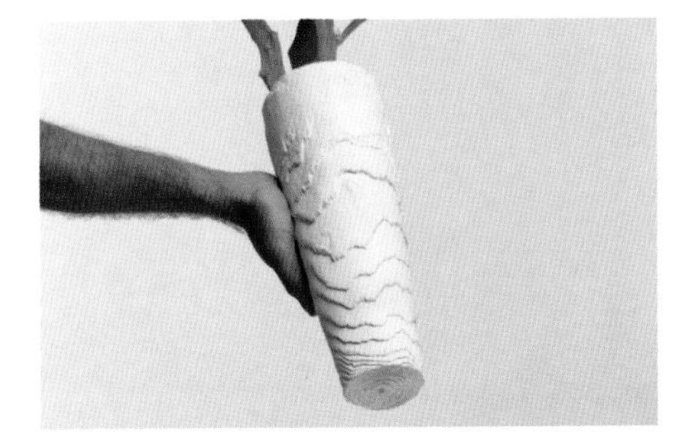

This process is versatile, as it can be used to smooth and shape objects, also for producing decoration similar to etching or carving.

Sandblasting does exactly what it says: abrasive particles are fed through a gun at very high speed, and effectively blast the workpiece. For obvious reasons, the safest way to use the process is inside a sealed chamber. In terms of decoration, sandblasting glass can be very effective and is considered to be something of an art form. It is possible to create all kinds of decorative patterns by using stencils, and simply adjusting the speed and angle of the propelled particles creates different shades, depths and effects.

Typical application
In addition to decoration, sandblasting is used to restore automotive parts, architectural structures and mechanical components, as it removes rust and corrosion. It has also been used to prime objects before painting. The small abrasive particles ensure that all imperfections are smoothed out and that dirt and dust are removed, which makes the paint stick better. One of the biggest uses has been for making denim jeans look distressed. However, inhaling the dust particles emitted during sandblasting can be hazardous. Legislation has been introduced to ensure factories operate with the appropriate set-up. In some cases the process has been banned.

Sustainability
The main alternatives to cleaning or altering the surface of a material are chemical based. Therefore, in comparison, sandblasting, which uses air (relatively low energy) and grit is more environmentally friendly.

Further information
www.lmblasting.com

Functional

i-SD System

i-SD is a new surface coating that provides an alternative to traditional hydrographics. It allows for highly accurate graphics to be draped over complex, 3D surfaces without distorting the image. The ink covers the entire shape, including cavities, recesses and detailed surface textures, from an original, high-resolution image. The process is compatible with many base materials, including plastics, wood, metal, glass and ceramic. With PBT plastic the image is actually embedded in the plastic so that it won't wear off.

Typical application
The i-SD process is currently being used in automotive applications where high wear-resistance is needed, but there is also a huge area of opportunity in mobile phone covers and decorative surfaces for other consumer electronics.

Sustainability
Through the combination with injection moulding the process results in faster production cycle times, resulting in less energy consumption than comparable decoration processes

Further information
www.idt-systems.com

In-Mould Decoration (Film Insert Moulding)

In-mould decoration was developed as an economical way of adding decorative surfaces to injection-moulded plastic parts. It eliminates the need for a separate printing process and is becoming more and more significant with the increased use of graphics and branding to personalise consumer products such as mobile phones and other electronics.

The process begins with printing the graphic onto a polycarbonate or polyester film, known as a 'foil', which is then cut to shape. Depending on the shape of the component to be moulded, the foil is fed into the mould as a ribbon, or cut into individually inserted pieces if the part is curved.

Typical application
In-mould decoration is not limited to text-based graphics, but can also be used to produce colour and surface patterns on mouldings. One of the most interesting, albeit invisible, foils that can be applied is a form of 'self-healing' skin which helps products stay shiny and free from scratches. Other applications include decorative mobile phone covers, digital watches, keypads and automotive trims.

Sustainability
In-mould decoration is more environmentally friendly than painting or spraying, which emit volatile organic compounds (VOCs).

Further information
www.macdermidautotype.com

Self-Healing Coating

This clear polyurethane coating has the ability to heal itself, repairing small scratches and blemishes on a product's surface, when exposed to heat. At high temperatures, the network of molecules in this plastic coating becomes elastic and it flexes, enabling it to smooth out scratches, similar to the way in which candle wax reacts when it is heated. The coating provides outstanding durability and resistance and could make the problem of scratches a thing of the past.

A good example of its use is for a car body. When the car is exposed to sunlight on a hot day small scratches on its sheet metal simply flow back together, improving smoothness and gloss.

Typical application
The coating has been tested for car-body sheet metal, but its possibilities are vast – perhaps even encompassing self- healing building surfaces?

Sustainability
This self-healing coating is environmentally friendly as it uses only a small amount of solvents.

Further information
www.research.bayer.com/en

Liquid-Repellent Coatings

Traditionally, liquid-repellent coatings rely on treating textiles and other materials with an impermeable coating, which generally completely alters the look and feel of the original material. P2i, on the other hand, produce a nanoscopic liquid-repellent coating based on plasma-enhanced vapour deposition technology. This uses a special pulsed plasma, operating in a vacuum chamber at room temperature, to polymerise a liquid-repellent monomer and attach it to the surface of the object being protected. The process forms a nanometer-thin, durable protective coating over every exposed surface of the object, making it completely liquid repellent, while leaving its other properties unchanged. The process works with a wide range of materials; even complex 3D objects incorporating several different materials can be treated successfully with the P2i process.

Typical application
The technology can be used to provide protection from liquids for a wide range of products, including electronics under P2i's Aridion™ brand and footwear under its ion-mask™ brand. Entire products, including seams and joints, can be coated in one go. Laboratory equipment and medical products are also important areas for the technology – for example, when applied to pipettes it ensures that all their liquid content is released for accurate test results.

Sustainability
The process requires only tiny quantities of protective monomer and waste is minimal, making it highly efficient compared to traditional methods such as dip application and spraying.

Further information
www.p2i.com

Ceramic Coating

Keronite® is a brand name for an extremely hard, wear-resistant ceramic coating that can be used with light metals and alloys, and has transformed surface engineering. It provides a more environmentally friendly, cost-effective and precise alternative to hard chroming and plasma spraying.

The application process starts with immersing the object in an electrolyte solution with an electric current passing through it. This creates a plasma discharge, which forms a thin primer layer in a process called 'plasma electrolytic oxidation' (PEO). Next is the functional hard layer, in which hard crystallites are packed into a crystal matrix that covers the entire part. Although the process is similar to anodising, it produces much thicker and harder layers while using less environmentally harmful alkali electrolytes.

Typical application

Keronite® has a unique combination of properties that makes it ideal for a wide variety of products. It has been approved for use on satellite hardware in the aerospace industry; the European Space Agency conducted a thermal shock test where Keronite® was immersed alternately in boiling water and liquid nitrogen to mimic conditions in space.

Benefits have been most recognised in the field of architecture. A light coating of Keronite® makes aluminium more robust and suitable for structural components. The image above shows a RockShox, 2011 Revelation World Cup adjustable bike fork.

Sustainability

The electrolytes used in this process do not contain environmentally damaging components and can be disposed of without treatment. Keronite® is 100 per cent recyclable.

Further information

www.powdertech.co.uk
www.keronite.com

Powder Coating

Powder coating is a completely dry process, the coating is made from a combination of finely ground resin, pigment particles and other raw materials, which are applied to the surface of an object. Tougher than conventional paint, it can be applied much more thickly without running or sagging. The process offers a choice of using a thermoplastic or thermoset polymer base material; the thermoplastic film will remelt upon heating, whereas the thermoset polymer will not change state once set.

Powder coating is most commonly applied by spraying, using an electrostatic charge to make the coating stick to the substrate. The object is first electrically grounded, then sprayed with a gun where the powder is passed by an electrode to charge the particles, causing an attraction to the grounded object. This electrostatic charge ensures an even layer of powder. After application, the object is placed in an oven where the powder particles melt and fuse to form a continuous surface.

Typical application

The toughness and robustness of powder coating make it ideal for demanding applications such as bicycle frames and automobile components, where scratching and weather conditions cause problems.

Because of the need to ground the substrate, powder coating was initially compatible only with electrically conductive materials, such as metals. However, there are various ways to get around this problem and it is now possible to coat other materials, including glass and MDF.

Sustainability

The process does not emit any volatile organic compounds (VOC) into the air. Additionally, unused or over-sprayed powder can be recovered and reused.

Further information

www.dt-powdercoating.co.uk

Phosphate Coatings

First developed at the turn of the twentieth century, phosphate coatings are still in common use today as a way of enhancing the properties of iron and steel parts. The coating acts a bit like a primer or base coat for metals, as it gives a great surface for painting while protecting against corrosion and wear.

As with many other finishes, the process begins with cleaning the component to be coated. Depending on its shape and size, the part is then placed on a rack, or in a basket or barrel, and immersed in a solution, where a thin compound of phosphate crystals forms over its entire surface.

Three types of phosphate are used. Zinc phosphate provides an excellent base for paints and has good anti-corrosive properties, whereas iron phosphate creates a good surface for bonding to other materials. Manganese phosphate is particularly effective at absorbing oil and also provides excellent resistance to wear.

Typical application
Phosphate coatings are used to prolong the life and minimise the maintenance of mechanical parts in a wide variety of applications, including automotive, aerospace and other heavy industries. They have also been shown to improve the biocompatibility of orthopaedic and dental implants, reducing the risk of rejection by the body.

Sustainability
The phosphate-coating process involves the use of harmful chemicals. However, the coatings provide excellent wear-resistance and corrosion protection, which prolongs the lifespan of product parts.

Further information
www.csprocessing.co.uk

Thermal Spray

This process is an extremely effective way of increasing a component's lifespan and performance. Although there are four types of thermal spraying, the fundamentals are very similar: a powder or wire is fed through a spray pistol, where it is heated until it is molten or soft and it is then sprayed onto the substrate. Thermal spraying can provide thick coatings over large areas where the density of the coating depends on the material used.

The different forms of the process are: flame, arc, plasma and high-velocity oxy fuel spraying, each of which uses various materials suited to a variety of applications. For resistance to atmospheric corrosion, thermal sprays are an excellent alternative to platings and paints, and have the added benefit that they are much less harmful to the environment.

Typical application
Thermal spraying has been remarkably successful in very demanding situations, such as electrical insulation for surgical scissors and improving the performance of bicycle brakes.

Due to its high cost, it is mainly used in the aerospace, automotive and biomedical industries, and for printing, electronics and food-processing equipment.

Sustainability
No volatile organic compounds (VOCs) are used, making thermal spraying an environmentally friendly process.

Further information
www.twi.co.uk

Case Hardening

Case hardening is a simple method used to harden mild steel. Heavy steels have a high carbon content and can be hardened by heating, whereas the carbon content in mild steel is too low for this. Instead, carbon is forced into the skin of the metal to produce a mild steel with an extremely hard outer surface, or casing, and a flexible and fairly soft core.

The process begins with heating the steel until it is red-hot. A part can be partially heated if only a smaller section needs to be hardened. The steel is then plunged into a carbon solution before being heated again and finally placed in cold clean water to cool. This process can be repeated to increase the depth and strength of the hardened surface.

Typical application
There are many applications for this process as it is suitable for all kinds of components that have to withstand high pressure and impact. Essentially, case hardening takes a material that is easy to shape – mild steel – and makes it very hard-wearing and durable. Treated parts cannot be cut with a saw and will not shatter easily.

Sustainability
Case hardening is not particularly efficient and it can be difficult to recover wasted material.

Further information
www.ttigroup.org.uk

High-Temperature Coatings

Diamonex is a thin but extremely hard-wearing coating with a diamond-like finish. It is normally possible to apply it at temperatures lower than 150°C, meaning Diamonex can be used with a wide range of materials, including plastics. In addition to its good wear- and abrasion-resistance, the coating is chemically inert and very hard with low friction.

Typical application
This incredibly versatile coating is used in everything from jet engines to supermarket checkout scanners – basically any application that requires super-tough wear-resistance and low friction. Diamonex is also suitable for many medical applications, including implants and surgical instruments.

Sustainability
Diamonex is an efficient coating process that doesn't leave much waste. However, it is worth bearing in mind that coated products are difficult to recycle and reprocess.

Further information
www.diamonex.com

Thick-Film Metallising

Thick film metallising makes it possible to 'print' a layer of metal onto plastics and ceramics. It can, in other words, be used to print fully functional conductive circuits directly onto the substrate, without the need for separate circuit boards. The coating can be applied by screen-printing, spraying and roller coating, and by using a laser where the metallised pattern is 'printed' directly onto the product.

Typical application
One of the most common uses for thick-film metallizing is in so-called RFID (radio frequency identification) tags. These are often used in the shipping industry to track parcels and other goods, and also in wireless ticketing systems, such as London's public transport Oyster card. But it could be used for so much more. Already, next generation rapid prototyping machinery has incorporated thick-film metallising technology, allowing designers to integrate working circuitry in their prototypes.

Sustainability
Because the metal is deposited directly onto the part, thick film metallising wastes only a minimal amount of materials. However, the metal must be removed to allow for plastic recycling of the product.

Further information
www.americanberyllia.com
www.cybershieldinc.com

Protective Coatings

The surface of glass is considered to be completely smooth, but on a microscopic level it is quite rough with tiny peaks and craters that cause dirt to stick to the surface. Diamon-Fusion® is a glass coating that improves the propoerties of glass while also protecting it. The coating fuses with glass to form a water-repellent barrier that also enhances visibility and strengthens the maximum weightload of the material by up to ten times that of untreated glass. Diamon-Fusion is also suitable for ceramics and most other silica-based materials, including porcelain and granite.

Diamon-Fusion® is applied using a process called 'chemical vapour deposition'. The surface to be treated is first cleaned and coated with a liquid catalyst. A special machine then emits a vapour containg the special chemicals needed to make the molecules change. The process happens inside a chamber that can be big enough to fit very large products. It takes only a short while and the glass can be used straight away.

Typical application
This versatile coating is used in glass and ceramic coating from bathroom fittings to car windscreens, where it can make a big difference in improving visibility and keeping the window clear during bad weather. Diamon-Fusion® also provides protection against damage from road debris, ice and snow, as well as acid rain and UV radiation. The coating is suitable for applications in marine environments.

Sustainability
Once the coating is applied to the glass it is chemically inert and completely non-toxic. The vapour deposition process used to create Diamon-Fusion® is also environmentally friendly. Additionally the use of the coating can lead to reduced cleaning cycles, which is good for the environment and uses less energy.

Further information
www.diamonfusion.com

Shot Peening

Shot peening is a process for cold-working metal surfaces to improve their strength and overall physical properties. To understand the process, think of a shotgun – shot peening basically involves pummelling a metal surface with lots of small round particles. The particles cause small dimples as they hit the surface, which creates a layer of highly stressed compression as the material beneath the surface tries to restore itself.

On a superficial level, shot peening is similar to sandblasting, but without being abrasive; this means that less material is removed during the process and that in some cases shot peening is even suitable for forming.

The process can in some cases increase fatigue life by up to ten times. As well as increasing strength, shot peening resists some forms of corrosion as it is difficult for cracks to form on the treated surface.

Typical application

Shot peening can be used for all kinds of applications – from architectural cladding to strengthening aircraft wings – where increased strength in metallic sheet materials is desirable. It is also sometimes used for forming, as opposed to just finishing, in aerospace industries. Additionally, the process can be used to strengthen materials after repairs.

Sustainability

Because shot peening is a cold-working process, it uses less energy than finishing processes that require heating. Unlike with sandblasting, little dust is created.

Further information

www.wheelabratorgroup.com

Plasma-Arc Spraying

Plasma is often referred to as the fourth state of matter. In the same way that sufficient cooling causes most materials to freeze, most solid materials turn into a plasma state when they are heated up enough. Plasma is very similar to a gas, but it has a unique property: it conducts electricity. Plasma arc spraying provides protection against high temperatures, corrosion, erosion and wear. It can also be used to replace worn material or to enhance the electrical properties of a material. The coating is compatible with a variety of base materials and can be produced in different thicknesses.

The spray material is usually a powder that is heated up and melted inside a spray gun. Once the material is molten, a gas that flows between an electrode and the nozzle is used to propel it to the work surface. Finally, as the material hits the surface it solidifies rapidly to form a solid coating.

Typical application

The resistance to high temperatures offered by plasma arc spraying is ideal for demanding applications in the aerospace industry, where a number of parts within turbine engines are sprayed so that they are able to perform in extreme conditions.

Medicine is another field where plasma arc spraying is hugely effective. The coatings are biocompatible, which allows a bond to be created between an implant and tissue.

Sustainability

Waste spray can be collected and reprocessed, making plasma-arc spraying an efficient process.

Further information

www.plasmathermalcoatings.com

Galvanising

One of the unique advantages of galvanising is its structure. The reaction between metals that the process causes, causes the coating actually to merge with the base metal to create outstanding toughness and enhance the longevity of metal parts.

A lot of preparation is involved in the process, as the part that will be coated must be completely and utterly clean if the reaction is to happen. So the part is first cleaned with a degreasing solution, then washed with water and placed in an acidic bath to remove rust and scale. Once it is entirely clean, it is dipped into molten zinc, which causes the zinc and the base metal to form a tough and inseparable protective layer all over the surface. The initial rate of reaction is very rapid and most of the thickness is formed at this time. The part is typically immersed in the molten zinc for about 4-5 minutes, but it can be for longer for larger products.

Typical application

The galvanising of steel parts is widely used within the construction industry. Steel bars, bolts, anchors, rods used in reinforced concrete, and motorway crash barriers are common applications that benefit from the increased durability and toughness the process offers.

Sustainability

Galvanising involves the use of some fairly nasty chemicals and acids, but the process is not a major environmental concern if it is managed properly.

Further information

www.wedge-galv.co.uk

Deburring

All machining processes – from shearing to drilling – inevitably cause untidy, rough and sharp edges on metals. These are called 'burrs' in the industry and deburring is the technique used to remove them.

There are a number of ways to remove the burrs, depending on the type of metal used and the shape of the product. The most common is to use a tumbler: the part is placed inside the drum along with small chips of various materials. It is then tumbled around inside the machine until all sharp edges have been ground down. The process also cleans, softens corners and sometimes even improves the strength of the part.

Typical application

Deburring is a critical step in the manufacture of parts for the aerospace industry. For example, the parts for a turbine engine will be subjected to extremely high pressures and temperatures during use, which means all the edges must be completely smooth with a generous radius. The process is simple enough to be used for post-cleaning any metal parts.

Sustainability

Automated deburring machines use a great deal of energy.

Further information

www.midlanddeburrandfinish.co.uk

Chemical Polishing
aka ElectroPolishing

Many components for electronic devices need to be extremely accurate geometrically and must have an equally high degree of surface finish. Chemical polishing achieves what can be very high tolerances in manufacturing, allowing for 'microscopically featureless' surface smoothness with minimal surface and structure damage.

The process works by exposing the component to controlled chemical dissolution in an acid bath. The acid attacks ridges and rough surfaces, causing them to dissolve faster than flat parts and leaving a perfectly smooth surface. If you are familiar with electroplating, think of electropolishing as the reverse, where metal ions are removed from a surface instead of being added to it.

Typical application
Chemical polishing is used for high-precision products such as electronics, jewellery, medical devices, razor blades and fountain pens.

Sustainability
The chemicals used in the process are aggressive but manageable, and excess material can be recycled.

Further information
www.logitech.com
www.electropolish.com
www.delstar.com

Vapour Metallising

Vapour metallising may not be the best known finishing process, but it has quickly become one of the most common ways to produce mirrors. It allows bright and reflective metal coatings to be applied, very cost-effectively, to a variety of base materials, including plastics. Vapour metallising can also be used as an alternative to electroplating in some applications to coat some parts of a surface and leave others uncoated.

The object to be coated is placed inside a jig and an adhesive base coat is applied in order to enhance the metallising process and ensure a durable coating. The base coat is cured in an oven, and the object is then placed in a vacuum chamber and evaporating aluminium (less common alternatives include nickel and chromium) forms an even coat all over the object. A protective topcoat is often applied as well.

Typical application
Because vapour-metallised parts are resistant to water corrosion, the process can be used for a number of car parts including wing mirrors, door handles and window trims. Kitchenware and bathroom fittings are common applications, as are metallic helium-filled party balloons.

Vapour metallising can also be used to give plastic materials a conductive metal coating. Packaging is another important area – just look at a packet of crisps to see an example of metallic coating of plastic film.

Sustainability
Vapour metallizing is more environmentally friendly than comparable processes such as electroplating, because it is cleaner and doesn't use toxic chemicals.

Further information
www.apmetalising.co.uk

Decallisation

Decallisation is used to apply photographic images, or any other graphics, to a wide range of substrates. It works by coating a base material with a layer of polyurethane, which is then printed, using screen- or offset printing, at temperatures of up to 200°C. The heat causes the ink and polyurethane to fuse, resulting in a very durable and scratch-resistant surface.

Decallisation's toughness, coupled with its ability to coat a wide range of materials, including plastics, metal, glass and MDF, makes the process extremely versatile, suitable for everything from architectural exteriors and interiors, transport and outdoor advertising to all kinds of consumer products.

Typical application
Decallisation is perfect for demanding architectural applications, such as bathrooms and kitchens – and even exteriors as it is UV-, abrasive- and graffiti-resistant. It is also suitable for high-maintenance areas such as public transport, stations and sports arenas.

Sustainability
Surfaces coated with this process cannot be recycled, but in many applications the toughness and durability it produces may lead to reductions in materials, maintenance and replacements.

Further information
www.decall.nl

Pickling

And no, this is not pickling as in preserving food. In this context, pickling is a method of cleaning various metallic surfaces. All kinds of manufacturing processes from cutting to welding can tarnish metals by leaving residue caused by oxidation, which discolours the surface. Before any additional layers, such as paints and coatings, can be applied to the metal the residue must be removed. This is where pickling comes into the picture.

The metal part is submerged in a bath of cleaning chemicals and heated up. It can take just a few minutes, or up to several hours, before the metal can be removed and washed. Large items can be sprayed with the chemicals, or a brush can be used to coat just specific areas.

By improving corrosion-resistance, pickling significantly increases the life-cycle of a product and improves its performance in use. It uses a variety of cleaning chemicals which are generally determined by the type of metal being processed. These include acids which remove a very tiny layer of the surface and therefore any scale.

Typical application
Pickling is often used in jewellery. Because the condition of the metal – which often involves copper, silver or gold – is important, any scale left after it has been soldered or fluxed needs to be removed from the surface. It is possible to purchase a pickle pot and process small items at home.

Sustainability
The waste products produced by pickling can be hazardous. However, the waste liquor can be reprocessed for the fertiliser industry. Alternatively, it can be recycled and used in the manufacture of steel.

Further information
www.anapol.co.uk

Non-Stick Coating (organic)

Based on plant cells, Xylan® is an organic range of fluoropolymer coatings that can improve the properties and usability of a wide range of materials. Like PTFE (more commonly known as Teflon®) Xylan is used for non-stick surfaces; the main difference is that it adheres very effectively to surfaces that would not usually accept PTFE.

The product to be coated is first degreased and cleaned so that the coating will stick properly. The coating is then applied in the form of a wet spray, which contains the fluoropolymer resin. The product is placed in an oven where the Xylan® cures and forms a thin film. The thickness of the coating depends upon the number of coats that are applied.

Typical application

Xylan®'s ability to increase the lifespan and performance of various components makes it particularly effective in the automotive industry. Here, aluminium is often used because of its low weight, but at the cost of the material's relatively scant durability. The use of Xylan® in this area helps aluminium to resist wear even in an environment of heat, oil and friction.

Sustainability

The lifespan and performance of a component is significantly increased, reducing the use of raw materials.

Further information

www.ashton-moore.co.uk

Non-Stick Coating (inorganic)

Teflon® is a brand that has captured the public imagination, to a degree that it is widely used, in all kinds of contexts, to describe a thing's, or even a person's, non-stick qualities. The scientific name for this plastic material is a mouthful – polytetrafluoroethylene, PTFE for short – so we should be thankful to the engineers at DuPont™ for coming up with its catchy brand name.

It is is extremely difficult to process this material using conventional methods for forming plastics, which is why PTFE is almost always used to coat other materials. The process works by spraying the substrate, then curing the coating in an oven where the PTFE forms a tough and uniform finish with remarkable properties: excellent self-lubrication and non-stick properties, and chemical- and heat-resistance. Other non-Teflon®, non-stick coatings include Xylan® (see right).

Typical application

Teflon® and PTFE are probably best known in connection with cookware, but the coating is also extensively used in textiles, such as GoreTex®, for improved weatherproofing. It is also used for medical equipment, where its heat- and chemical-resistance help to maintain exacting standards of cleanliness and sterility.

Sustainability

PTFE and in particular one of its ingredients – PFOA – is often said to be a potential threat to the environment. It is worth noting that DuPont has recently removed PFOA from the Teflon® production process and that the US Environmental Protection Agency does not advise against normal use of non-stick cookware and PTFE-coated all-weather clothing.

Further information

www.dupont.com

Decorative & Functional

Chrome Plating

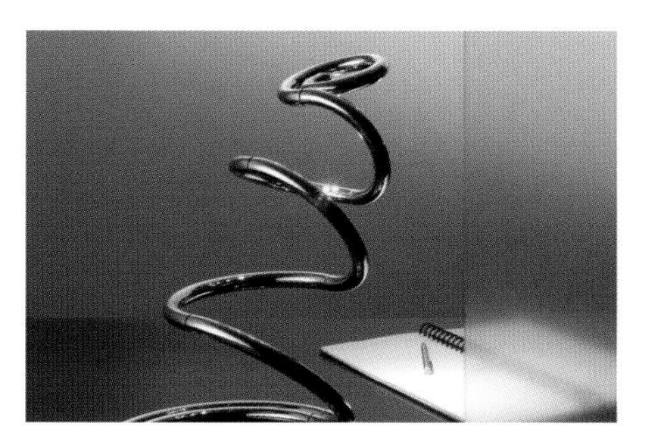

Chrome plating is a technique commonly used for coating objects that require particular resistance to corrosion and wear. There are typically two types of chrome coating, the most common of which is the thin, decorative, bright chrome that can be used on a wide range of products, followed by hard chromium plating, which is much thicker and is often used on industrial equipment to reduce friction and wear.

The component must be thoroughly cleaned and buffed in order to create a smooth, even surface. It is then electrically charged and immersed in the chromium solution which is also charged. The charges cause an attraction between the surface of the component and the solution which produces an even layer over the entire surface of the object.

Typical application

Chrome plating first became mainstream within the automotive industry for detailing such as bumpers, handles and mirrors, as chrome has excellent resistance to corrosion.

Bathroom fixtures are another important area where chrome plating is applied, it is suitable in other applications where moist and humid conditions are common. It is also used in purely decorative applications such as the Pizza Kobra light by Ron Arad.

Sustainability

Chrome is difficult to recycle as some compounds are toxic and can have a damaging effect on the environment. Although the production of chromium can release harmful emissions, the process has improved environmentally since 1970.

Further information

www.advancedplating.com

Anodising

One of the most interesting facts about this finish is that it is a protective skin that is grown from aluminium, toughening and thickening the natural oxide that is contained in the metal. The components to be anodised are thoroughly cleaned before being immersed in a sulphur solution. A current is passed through the aluminium component and converts the aluminium surface to form the aluminium oxide layer. The thickness and hardness of the coating are determined by the strength of the electrical current, the temperature of the sulphur solution and the length of time the component is immersed in the solution. There are various forms of anodising chosen according to whether the main requirement is production or decoration. Although aluminium is the main metal used for anodising, titanium and magnesium can also be anodised; just think of the metals as 'the three -ums'.

Typical application

Apple's Mini and Shuffle iPods used anodising to create a tough and protective coating for the aluminium casing, while offering a variety of seductive colours. Another design icon is the Maglite® torch (p.18), which uses anodising to communicate an industrial aesthetic. The low weight of aluminium and the durability and corrosion-resistance of anodising is a perfect combination for this type of application.

Sustainability

More environmentally friendly than many of the other metal finishing processes, anodising releases fewer toxins in comparison. An anodised finish is non-toxic and the chemical baths used in the process are often reclaimed, recycled and reused.

Further information

www.anodizing.org

Decorative & Functional

Shrink-Wrap Sleeve

Shrink-wrap sleeves are used as a protective layer on a vast range of products and are encountered on a daily basis. The sleeve is made of a thin plastic film that shrinks tightly and encapsulates a product when heated. The shrinking occurs because the film is manufactured so that the molecules are arranged in a random order. Heating the film causes the molecules to set and so reduces the size of the film. Shrink-wrap sleeves are available in a variety of thicknesses, clarities, strengths and shrink ratios. Sleeves can be made to shrink in one direction (monodirectional) or both directions (bidirectional).

The film can be printed, which offers excellent possibilities for branding and other graphics. It is often easier to print graphics onto a shrink-wrap sleeve rather than the primary packaging.

Typical application
Shrink-wrap sleeves are commonly used to overwrap many types of packaging, including drinks cans and bottles, CDs and DVDs, cartons, books and even whole pallet loads. They can also be used as a primary covering for foods like cheese and meats.

Sustainability
Shrink-wrap sleeves can be recycled together with other plastics.

Further information
www.sealitinc.com

Dip Coating

The process of dip coating is similar to dip moulding, but there is one major difference between them: with dip moulding a plastic object is produced and the mould is removed, whereas with dip coating a permanent plastic layer is created over an object made of another material, usually metal. Dip coating provides a very stable and protective coating that is often decorative, and also ergonomic in improving grip on products such as handles.

The process begins with heating the object to be coated, then placing it in a container and blowing plastic powder over it from all directions to create an even layer. The heat from the object causes the plastic powder to melt and stick to the surface. The coated object is then returned to the oven and reheated until the plastic layer is completely smooth, when the object can be taken out and left to dry.

Typical application
Dip coating is ideal for grips on hand tools such as pliers and clippers, as the plastic coating provides a softer and more comfortable grip than the substrate. Other applications include outdoor furniture, automotive clips and fitness equipment.

Sustainability
Dip coating can be energy efficient in longer production runs, when a large number of products can be coated at the same time.

Further information
www.omnikote.co.uk

Ceramic Glazing

Because ceramic materials are porous, most products made from them would be unable to hold liquids without a layer of glazing. The glaze gives ceramics a glass-like surface that is impermeable and protects any surface decoration underneath.

To apply the glazing a dry powder is dusted over the ceramic object using an airbrush, or the object can be dipped into the powder. The object is then fired in a kiln, which causes the powder to soften and flow over the ceramic surface. A reaction between the ceramic and the powder causes a strong bond between the two. It is worth noting, though, that the part of the product that will be in contact with the kiln must be left unglazed, otherwise it will stick to the kiln – if you have ever wondered why the base of a tea cup has a different texture to the rest of it, this is why.

Typical application

Ceramic glazes have been used for thousands of years for all kinds of ceramic products. They continue to be used today for cookware, plant pots, storage containers and thousands of other applications.

Sustainability

Glazing significantly increases the lifespan of ceramic products by producing a strong, durable and water-resistant coating. The main environmental issue is the amount of heat required to fire them.

Further information

http://glasstechnologys.com/

Vitreous Enamelling

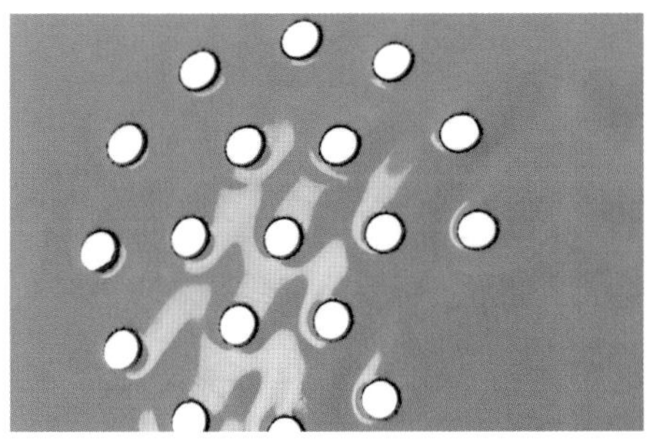

Enamelling has been used for thousands of years for its decorative and protective properties. It is essentially a sophisticated process that uses heat to fuse a thin layer of glass powder onto a metal surface. Various colours can be produced using different types of mineral.

The metallic surface of the object to be enamelled is first engraved with the desired pattern or shape. Powdered glass is then carefully poured into the grooves of the engraved shape and the object is fired in a kiln. The heat melts the powdered glass and the resulting liquid spreads evenly within the shape. When the object cools down the enamel hardens to form a hard, smooth, glass surface.

Typical application

Enamel is resistant to heat and wear, and is therefore often used in everyday products such as kitchen hobs, saucepans and washing-machine drums.

Because the coating is completely fireproof and the colours remain vibrant for hundreds of years, vitreous enamelling is useful for signs and other graphics – for example, the famous station signs and maps for the London Underground.

Sustainability

Enamelled products are extremely durable, and the brilliance of the original colours is still visible after hundreds of years. The main issue is the amount of heat required to fire them in the kiln.

Further information

www.kingfisherenamelling.com

Glossary

annealing The process of reducing the stresses in a material by the application of heat in a controlled manner. It involves glass or metal being heated and/or cooled slowly in a lehr (an oven or kiln), thus bringing about a relaxing of the internal stresses in the material.

axisymmetric A three-dimensional form that is symmetric around a single axis; a typical example might be a cone.

bar A term used in industry to describe pressure in a vessel. 1 bar is equal to 14.504 pounds per square inch (psi), 0.98692 atmospheric pressure or 100,000 pascals (Pa).

billet In engineering terms a billet is a solid lump of steel from which are made rods, bars and sections.

biscuit/bisque Refers to a ceramic that has been fired but not glazed. The firing takes place at around 1000°C. A ceramic piece that is biscuit-fired is porous.

burring A rough, often sharp, edge left on a piece of metal after it has been cut, cast or drilled. Although only a minor by-product of various forms of production, it is sufficiently noticeable that companies have been set up expressly to deal with deburring in its many forms.

CAD Computer-Aided Design.

chip-forming cutting Methods of making components that create chips of material as a result of the cutting process. A typical example would be milling. *See also* non-chip-forming.

CNC Computer Numerical Control.

CNC folding A process used to create three-dimensional hollow forms by the action of computer-controlled creasing and folding of a flat sheet material. Think of a child's metal pencil-case.

cold working Working and forming metal or glass at a temperature below that at which it recrystallises, or, in simple terms, without the use of heat. *See also* work hardening.

commodity polymers These have a lower mechanical performance than engineering polymers (see below) and include polypropylene and polyethylene.

composites Materials that are made of two or more ingredients. The term is generally used to refer to materials with advanced properties that are made from a combination of polymer resins and fibres.

crazing An imperfection in a ceramic glaze that appears as a fine cracking.

deep draw Refers to a component that has been produced by a long punch that draws the metal into a deep shaft. Impact extrusion is an example of a process that produces a deep draw.

die The terms 'die' and 'mould' are virtually the same and refer to a form, generally made of steel, that is used as a cavity into which material is added, and which imparts its shape onto the component. *See also* tools.

draft angle *See* draw.

draw This is the taper that you will need to consider when designing parts using many moulding processes. It is generally a slight angle that allows parts to be easily ejected from a mould.

engineering polymers Plastic materials with performance characteristics, for example: nylon, acetyls and thermoplastic elastomers (TPEs). *Compare* commodity polymers.

fabrication A metalworking term used to describe the construction of components by assembling and fixing together various parts, rather than the manufacture by, for example, moulding or casting in a single operation.

fettling A final cleaning-up of a ceramic piece before it is fired.

flange A lipped detail or rim that is usually straight and located on the edge of a metal part. Its function is to add stiffness and to facilitate joining to another part.

flashing In production, the flashing (or flash) is the excess metal left on a part after a forming operation. It is unwanted and generally needs to be removed.

gate This is a term that is mentioned quite often in connection with plastic moulding. It refers to the orifice through which the hot, molten plastic enters the mould cavity.

gel coats A term specific to composites, it refers to a quick-setting coating applied to the internal surface of a mould to provide an improved, highly glossy, protective surface finish.

gob A term used in glass blow moulding that refers to the sausage-shaped, measured quantity of molten glass before it enters the mould and is blown.

'green' state The wet, semi-hard physical state of a ceramic component before it has been fired.

jig In production, a jig is a structure used to control or restrict the movement of a component or material while it is being worked on, assembled or glued.

lathing Otherwise known as turning but generally used in the context of working in metal.

mandrel Often used in metal spinning to describe the solid form against which the sheet of metal is spun to achieve the desired shape.

matrix When using composites, the matrix is the material to which the fibres are added – often a liquid polymer.

micron One thousandth of a millimetre.

mould *See* die.

non-chip-forming cutting
A term used to describe cutting methods that do not result in chips of material being formed. These methods are very clean and include, for example, laser or water-jet cutting, where the material is blasted or vaporised leaving no 'chips'.

outgasing A term used to describe the emission of a volatile gas during the processing of plastics, for example in injection moulding. There are many established ways of removing these gases.

parting lines The fine line that stands proud on the surface of a component that is often left after moulding. Essentially it is where the two or more parts of a mould have separated. Also known as witness lines.

post-process operations Any process that takes place after the main production is referred to as 'post'. Examples include: post finishing, where surfaces might need to be cleaned up; and post forming, post working or post-machining where a secondary process is used to complete the component. These might include drilling a hole or deburring.

pre-form Used predominantly in blow moulding to describe an injection moulded, semi-formed component before it has been fully moulded. A kind of product in its embryonic stage.

psi A unit of pressure that stands for pounds per square inch. *See also* bar.

re-entrant angles
See undercuts.

refractory materials Materials with very high temperature-resistance that are used in furnaces or kilns. Many ceramics are 'refractories', which is another way of describing these materials.

risers A term used in metal casting to describe a shaft in the mould that acts as a reservoir from which molten metal can be drawn to offset the shrinkage that takes place in metal casting once the metal cools and solidifies. *See also* runners.

runners The shaft into which metal is poured during casting. *See also* risers.

sink marking This is a common, often easily solved, problem that occurs in injection moulding. It is when plastic that needs to be formed into a flat surface exhibits a slight indent or depression, the 'sinking'. This is often because of local shrinkage of the material within the component.

solid-state forming An umbrella term used to describe the processing of materials usually at room temperature. Examples include impact extrusion and rotary swaging.

sprue This is the tapered piece of plastic that is left attached to a component as the result of injection or compression moulding. It occurs where the plastic flowed into the mould from the nozzle. Evidence of where this has been cut off can often be seen in cheap mouldings.

substrate A term generally used to describe the surface material onto which a secondary layer of material is applied. It can be considered as something like the base material.

tempering The purpose of tempering is to reduce the hardness of steel by relieving the stresses in the material. The process involves heating the steel to a temperature below the transformation range and then cooling it slowly in air.

tensile strength When a material is stretched so that the length increases and the cross-section decreases, tensile strength is the amount of stress that the material is able to withstand.

thermal expansion Most materials expand when heated and contract when cooled. Thermal expansion is defined as the ratio between increase in temperature and increase in dimension of the material.

thermoplastics Together with thermosets (see below) the term is one of the main classifications for grouping plastics. Unlike thermosets, thermoform plastics can be reheated and subsequently re-moulded.

thermosets Plastics that once formed cannot be re-heated or re-moulded. Also known as thermosetting plastics. Compare thermoplastics.

tools A general term (also 'tooling') that refers to the part of the production set-up that could loosely be described as a mould (see above). However, tooling is not necessarily the male or female mould itself, but the complete mechanism that is in direct contact with the material. It may refer to either a mould, cutters or formers.

undercuts This is a very common term used in many forms of moulding to describe details in a component that would restrict that component's removal from a mould. Also often referred to as re-entrant angles.

witness lines *See* parting lines.

work hardening The best way of explaining this phenomenon is that when a piece of metal is bent ('worked') continuously it becomes increasingly hard and difficult to bend, and eventually it breaks. Annealing (see above) at regular intervals prevents this and allows for further working. See also cold working.

Index

Acknowledgements

Thanks to Dani Salvadori, who has helped champion this idea since its inception. Many thanks also to Ishbel Neat, Lucy Macmillan and Jennifer Hudson for their help with the laborious task of chasing permissions for the images. As always, a massive thank you to 'my team' for this project: Hema Vyas, Hayley Ho, Anna Frohm and the biggest thank you to Daniel Liden, who has been many things in the writing of this book, including researcher and adviser. Thanks also to James Graham for his wonderful illustrations; to Roger Fawcett-Tang for making this book so visually striking; Xavier Young not only for his photography but also for his inspired ideas and continued collaboration; to Russell Marshal for his technical appraisal; and to Alan Baker who, after almost 20 years, is still my 'materials and process man' – it is a pleasure to have been one of your students.

A really big thank you to all at Laurence King Publishing, particularly Jo Lightfoot and Jessica Spencer.

A special thanks to my young sons Theo and Jerome, whose constant joy provided me with the motivation to deal with the frequent lack of sleep they both so generously encouraged. Lastly, thank you to my wife Alison who after 20 years continues to be my supreme creative inspiration, partner and motivator. Here's to the next 20 years, my love.

440369

Picture Credits

The author and publisher would like to thank all contributors who have kindly provided images for use in this book. Every effort has been made to contact copyright holders, but the publishers would be pleased to correct any errors or omissions in any subsequent edition.

All illustrations by James Graham

Photographs by Chris Lefteri: pp. 20, 28, 31, 32, 66, 104 left, 110, 142, 206

Photographs by Xavier Young: 18, 24, 29, 40, 50, 64, 99, 108–09, 120, 130 top, 134 right, 140, 149, 170, 172, 174, 187, 209, 219, 231, 236, 275

18 By permission of Mag Instrument, Inc. MAG-LITE and MAGLITE are registered trademarks of Mag Instrument, Inc.
21–22 DEMAKERSVAN
24 Courtesy of Arcam AB.
26 Wade Ceramics Ltd.
29 WWRD UK LTD.
33 Reproduced by permission of TWI Ltd.
38 left Sam Buxton, Mikro Man Off Road, 2002, 95lx35wx40h mm. Image courtesy of the artist.
38 right Sam Buxton, Mikro Man Player, 2002, 95lx35wx40h mm. Image courtesy of the artist.
40 Normann Copenhagen.
42 Design: Louise Campbell. Manufacturer: Hay
46 Fuminari Yoshitsugu
50 Acme WhistleCo.
52 Setsu & Shinobu ITO. FIAM Italia Spa.
56 Peter Mallet
58 Heatherwick Studio
59 Rexam Beverage Can Europe & Asia
61 & 62 Courtesy of Industrial Origami, Inc.
67 & 68 Courtesy 3D-Metal Forming B.V.
70 Image Marc Newson
73 & 74 All images Zieta Prozessdesign
76 Created by Stephen Newby/ Full Blown Metals. Image taken by Joe Hutt.
78 Courtesy of Sodra Cell International AB.
80 Lapalma s.r.l., via E. Majorana 26, 35010 Cadoneghe (PD), Italy.
83 Design: Komplot Design/ Poul Christiansen & Boris Berlin. Production: Gubi, Denmark.
85 Photographs by Reholz GmbH.
86 right Neville UK Plc.
94 Photograph by Ida Riveros
96 Heatherwick Studio
97 Peter Mallet
99 © Exel Composites Plc.
101 Photographs by Daniel Liden, taken at RBJ Plastics in Rickmansworth, www.rbjplastics.com
104 right Courtesy of Apple Inc. Photographer: Doug Rosa
108-9. Permission for LKP to use image has been granted by Potter & Soar Ltd. who are part of the Aughey Group of companies.
112 Designer: Antoni Arola. Photograph by Carme Masià.
116 left Mathmos Airswitch Flask Lamp is a trademark and a patented technology owned by Mathmos – www.mathmos.com.
116 right Courtesy of Kosta Boda. Photographer Vassilis Theodorou
118 left Photograph by Goran Tacevski. The vases were developed with Mr. Karel Krajc, Head of the Technology Department of glassworks at Kavalier Sázava.
118 right Photographs by Craig Martin and Brian Godsman, property of Scott Glass Ltd.
120 Courtesy of KIKKOMAN CORPORATION.
123 Photographs courtesy of Beatson Clark Ltd.
124 Bruni Glass SPA copyright. Photograph by Bruni Glass SPA.
130 bottom Courtesy Marcel Wanders
134 left Wade Ceramics Ltd.
135 Courtesy of Diptech.
137 & 138 Courtesy Marloes ten Bhömer
143 & 144 top. Courtesy of Darmstadt University of Technology
146 & 147. Sigg Switzerland AG
149 Courtesy of Vernacare
154 Courtesy of Polyworx. Infusion of the Southernwind 100' carbon/epoxy sailing yacht; technology designed and implemented by Polyworx BV.
158 Courtesy of Mathias Bengtsson.
159 Photographs courtesy of Goodrich Crompton Technology Group.
170 © KYOCERA.
172 NGK Spark Plugs (UK) Ltd.
176 By permission of Go-Ahead London
179 Seggiolina Pop, Magis, photo by Carlo Lavatori. Magis SPA, Z.I. Ponte Tezze – Via Triestina Accesso E, 30020 Torre di Mosto (VE), Italy. Tel: +39 0421 319600, Fax: +39 0421 319700, info@magisdesign.com
181 & 183. Photographs courtesy of Alias
184. & 186. Photographs courtesy of Malcolm Jordan.
192 Fraunhofer Institute Material and Beam Technology IWS, Germany.
196 Courtesy of BIC® Cristal®
201 Air-chair designed by Jasper Morrison. Air-chair, Magis, photo by Tom Vack. Magis SPA, Z.I. Ponte Tezze – Via Triestina Accesso E, 30020 Torre di Mosto (VE), Italy. Tel: +39 0421 319600, Fax: +39 0421 319700, info@magisdesign.com.
203 Courtesy Gianni di Liberto
206 Stanley UK Sales Ltd.
212 Courtesy IDT Systems
214 Courtesy of Tulip Computers.
216 Courtesy of Metal Injection Mouldings Ltd.
219 Courtesy Mattel. LOTUS, EUROPA and the Europa car design are the intellectual property of Group Lotus plc. N.B. The rights of Group Lotus plc extend to the design of the car embodied in the toy featured in work and the trade marks LOTUS and EUROPA. Group Lotus owns no rights in the toy itself nor in the photographs of the toy to be reproduced in the book.
222 Photograph Ida Riveros
224 Reproduced with permission of Rolls-Royce Motor Cars Limited. The Spirit of Ecstasy, Rolls-Royce name and logo are registered trade marks and are owned by Rolls-Royce Motor Cars Limited or used under licence in some jurisdictions. Photograph by permission of Polycast.
226 Photographs by permission of Polycast.
228 Courtesy of Olof Kolte
234 Marc Newson Ltd
236 WWRD UK LTD.
240 Homaro Cantu/ Moto Restaurant, Chicago
242 Models produced by Mcor on the Matrix 300 file provided by Paul Hermon Queen's University and photographed by Cormac Hawley
244 Photographs courtesy of the Centre for Rapid Automated Fabrication Technologies (CRAFT), University of Southern California.
246 Black_Honey.MGX by Arik Levy for .MGX by Materialise.
248 top left/right Patrick Jouin ID – Solid C1 – Patrick Jouin studio.
248 bottom left Patrick Jouin ID – Solid C1 – Patrick Jouin studio.
248 bottom right Patrick Jouin ID – Solid C1 – Thomas Duval.
250 Courtesy of Mimotec.
253 Photograph courtesy of Renishaw plc – www.renishaw.com
255 Courtesy of Cornerstone Research Group.
257 & 258 Photograph by Martin McBrien. Courtesy of Dr Julian Allwood and University of Cambridge Dept. of Engineering
262 (Sublimation Dye Printing) Alessandro Paderni; (Vacuum Metallizing) left: Courtesy Tom Dixon right: Ida Riveros
263 (Flocking) Courtesy Miele & Cie KG; (Acid Etching) Courtesy Studio Tord Boontje
265 Courtesy Franc Fernandez and kxdesigners
267 Adrian Niessler and Kai Linke
269 (Self-healing coating) Courtesy Hyundai; (Liquid-Repellent Coatings) Courtesy p2i
270 (Ceramic Coating) Courtesy of SRAM LLC; (Powder Coating) Courtesy COLNAGO
274 Courtesy WHEELABRATOR GROUP
279 (Chrome Plating) Amendolagine e Barracchia Fotografi; (Anodizing) Image courtesy of VERTU, pioneer of luxury mobile phones
281 (Ceramic Glazing) Courtesy Porzellan Manufaktur Nymphenburg www.nymphenburg.com; (Vitreous Enamelling) Xavier Young